OVERFISHING

WHAT EVERYONE NEEDS TO KNOW

RAY HILBORN, WITH ULRIKE HILBORN

OXFORD

UNIVERSITY PRESS

OXFORD
UNIVERSITY PRESS

Oxford University Press, Inc., publishes works that further
Oxford University's objective of excellence
in research, scholarship, and education.

Oxford   New York
Auckland   Cape Town   Dar es Salaam   Hong Kong   Karachi
Kuala Lumpur   Madrid   Melbourne   Mexico City   Nairobi
New Delhi   Shanghai   Taipei   Toronto

With offices in
Argentina   Austria   Brazil   Chile   Czech Republic   France   Greece
Guatemala   Hungary   Italy   Japan   Poland   Portugal   Singapore
South Korea   Switzerland   Thailand   Turkey   Ukraine   Vietnam

Published by Oxford University Press, Inc.
198 Madison Avenue, New York, NY 10016

www.oup.com

Oxford is a registered trademark of Oxford University Press

Library of Congress Cataloging-in-Publication Data
Hilborn, Ray, 1947–
Overfishing : what everyone needs to know / Ray Hilborn with Ulrike Hilborn.
p.   cm.
Includes bibliographical references and index.
ISBN 978-0-19-979813-1 (hardcover) — ISBN 978-0-19-979814-8 (pbk.)
1. Overfishing.   2. Sustainable fisheries.
3. Fisheries—Environmental aspects.   I. Hilborn, Ulrike.   II. Title.
SH329.O94H55 2012
338.3'727—dc23   2011031308

1 3 5 7 9 8 6 4 2

Printed in the United States of America
on acid-free paper

To Carl Walters, whose curiosity and creativity have provided constant inspiration

CONTENTS

9 Deepwater Fisheries   69

10 Recreational Fisheries   78

15 Ecosystem Impacts of Fishing   110

16 The Status of Overfishing   122

PREFACE

On November 3, 2006, the *New York Times* ran a front-page article reporting that current fish stocks were on their way to collapse. The story, apocalyptically titled "Study Sees 'Global Collapse' of Fish Species," cited expert predictions that if "fishing around the world continues at its present pace, more and more species will vanish, marine ecosystems will unravel and there will be 'global collapse' of all species currently fished, possibly as soon as midcentury." Based on a press release that focused on one paragraph in an otherwise doom-free paper published in *Science*, the most prestigious scientific journal in the United States, the story went global, hitting the front pages of most major newspapers and making the BBC evening news. This particular story has had astonishing staying power but is only one of many about the demise of world fisheries and the collapse of marine ecosystems that has circulated in the last 10 years.

Yet in 2009, several of the same authors of the 2006 study concluded in another *Science* paper, entitled "Rebuilding Global Fisheries," that, after studying the trends in abundance and the percentages harvested for 167 fish stocks from around the world, "the average exploitation rate... is now at or below the rate predicted to achieve maximum sustainable yield for seven [out of 10] systems." Unsurprisingly, there were no global headlines.

And the contradictions continued. Two months after publication of "Rebuilding Global Fisheries," an article appeared in *The New Republic* entitled "Aquacalypse Now: the End of Fish," by Daniel Pauly, arguably the best-known fisheries scientist in the world. In 2010 we had news that cod in the North Sea and the Baltic Sea, both considered on the verge of collapse by many, were actually rebuilding and the World Wildlife Fund, an NGO active in marine conservation, put North Sea Cod back on the menu. More good news came in early 2011 when Steve Murawski from the University of South Florida and former chief fisheries scientist for the U.S. government announced that overfishing had ended in the United States.

The public can be forgiven for being confused.

So what's the story? Is overfishing killing off ocean ecosystems or are fisheries being sustainably managed?

It all depends on where you look. There are enough horror stories about the collapse of fisheries to fill volumes, and those volumes have been filled. *The End of the Line, Sea of Slaughter, Ocean's End,* and *The Unnatural History of the Sea* all tell stories of overfishing and the plundering of marine resources.

Aside from such rape and pillage, commercial fishing has suddenly and somewhat inexplicably begun to hold the viewing and reading public's interest. Linda Greenlaw became something of a cult hero with her book on swordfish fishing titled *The Hungry Ocean: A Sword-Boat Captain's Journey,* followed by the television series *Most Dangerous Catch,* which brought the daily lives and perils of commercial fishing into millions of homes without dwelling on any environmental aspects.

The devil, as always, lies in the details. Overfishing is too complex a story to be told in a clean beginning-middle-and-end kind of narrative.

Let's look at the response to the 2006 paper suggesting that all fish stocks could be gone by 2048. My fisheries experience up until that point had largely been on the west coasts of the United States and Canada and in New Zealand. Alaska and

New Zealand in particular had been widely considered to have some of the best-managed fisheries in the world, and on the west coast of the lower 48, overfishing had been greatly reduced and formerly depleted stocks were rebuilding. I knew that these fisheries, at least, were not collapsing and therefore all fish would not be gone by 2048. Because of this comment, the U.S. National Public Radio invited me and Boris Worm, the lead author, to have it out.

Boris Worm is a young professor at Dalhousie University in Canada. He grew up in Germany and had seen the decline in marine ecosystems in both Canada and Europe, a very different experience from mine. After the broadcast, Boris and I began a conversation exploring why we had such diverging perspectives on the sustainability of world fisheries.

The projection that all fish would be gone by midcentury was based on an examination of the catches of individual stocks, with the assumption that if the catch of an individual fish stock declines to less than 10% of its previous maximum, the fishery has "collapsed." If you plot the proportion of world fisheries that were thus deemed collapsed and project an accelerating trend forward, 100% of all stocks would indeed seem to collapse by 2048.

Boris and I agreed that catch is not necessarily a good measure of the actual abundance of fish stocks, and we initiated a joint study with 19 other scientists who work on marine fisheries to assemble all the estimates of actual abundance we could find.

Fish abundance is often measured by scientifically designed surveys, so we compiled a database with all the survey information publicly available. Many fisheries agencies around the world also use surveys in addition to other information to calculate historical trends in abundance, catch, and percentage of the population harvested. This analysis is called "stock assessment," and we assembled a different database with all the stock assessments we could find. When we wrote the 2009 paper in *Science* there were almost 200 fish stocks in that data set. The work continues and as of January 2011 we have reached 300 stocks.

We called our project "Finding common ground in marine conservation and management" and in the end all of us stood on that patch of common ground. We confirmed that about two thirds of the stocks for which we had data were at population sizes lower than the targets set by national and international agencies, and that the number of stocks at low enough abundance to be called "collapsed" was growing. We also found that fishing pressure had been reduced in most of the places we studied, and that most fish stocks were now fished at rates that would lead to rebuilding, not collapse. We also found that the overall trend in fish stock abundance was not downward but stable.

This group of 21 authors comes from a range of backgrounds, geographic regions, and pre-existing perspectives, but once we looked at actual abundance of fish stocks we had little trouble writing a paper that laid out what we had found. My own experience that Alaska and New Zealand had somehow avoided overfishing was confirmed. The fisheries off the west coast of the lower 48 states were indeed rebuilding. Boris's experience, too, was confirmed—in eastern Canada and most of Europe overfishing had been a major problem and stocks were often well below target levels. The data really speak for themselves. The most important finding, however, was that fishing pressure, the driver of overfishing and collapse, was generally being reduced.

The paper has been criticized for a bias toward Europe and North America. At the time we had almost no data from Asia, Africa, and South America, and those places are still underrepresented even though our database continues to expand. However, we do know from other studies by the Food and Agriculture Organization of the United Nations that overfishing has been more of a problem in the North Atlantic than anywhere else in the world, and that was the focus of our study.

At the same time, progress has been made in the North Atlantic to stop overfishing and reduce exploitation. This is not necessarily true for the areas for which we lack data. To them the hopeful message of our 2009 paper may not apply.

Again, the story of overfishing is not simple and certainly not the same everywhere.

There are places that have been severely overfished and others that have not. Some management agencies have reduced fishing pressure and stocks are rebuilding, while elsewhere fishing pressure has been left too high and over-fishing continues.

A dedicated writer could certainly pick through our data set for a book on overfishing and collapse whereas a different one could pick differently and fill a book with great successes.

In this book I have attempted to tell the stories of over-fishing and sustainable fishing, of failures and successes in fisheries management and hope to guide you, as impartially as I can, through the scientific, political, and ethical issues of harvesting fish from the ocean.

Fish are not the center of our understanding of fisheries. There is a wide web of intricate relationships between marine ecosystems and what we take from them, the people who catch fish, the social and economic fabric of communities and markets, and the governmental institutions that regulate the fisheries. To maintain sustainable fisheries we must also maintain sustainable ecosystems, sustainable communities, and sustainable economic activity.

If the fish were indeed at the center, we could simply stop fishing. The consequences, though, would be dire. Countless fishing communities around the world, the very reason fisheries exist, would have their livelihood and social fabric destroyed. And we would have to think hard about how to replace the 25%, or one quarter, of animal protein that fish provide on dinner tables worldwide.

The demand for food will steadily increase with a growing human population and we must recognize that fish from the ocean are a major source of sustainable protein. Lock up the oceans to fishing and there will be worldwide food shortages.

The oceans are unique in being able to provide large amounts of food from natural ecosystems. When sustainably managed, a marine ecosystem retains its structure and function despite major changes. Yes, the abundance of fish will be reduced by more than half and there will be fewer large, old fish, but the same species will be there in a still-wild ecosystem. Contrast that with agriculture, whose first steps involve cutting down or plowing up a native ecosystem and replacing natural species with exotic ones.

There are many reasons to avoid overfishing—world food security depends on it, marine birds and mammals depend on it, and employment for millions of people depends on it.

I hope this book contributes to the sustainable use of the oceans.

Ray Hilborn
Seattle, March 2011

A note on the use of *fisherman*. Because there is no collective noun in English to encompass the men and women who fish for a living or for sport and I have been severely chastised for using "fisherwoman" or "fisher" (nor do I feel qualified to invent a new word), I have taken the easy way out and used fisherman throughout. Please consider it all-inclusive and be assured that no slight is intended.

OVERFISHING

WHAT EVERYONE NEEDS TO KNOW

1
OVERFISHING

What is overfishing?

Overfishing is harvesting a fish stock so hard that much of the potential food and wealth will largely slip through our fingers. Yield overfishing is the most common. It prevents a population from producing as much *sustainable yield* as it could if less intensively fished. The population will typically be less abundant, but it can and often does stabilize in an overfished state. However, with extreme overfishing, in which the forces of decline are consistently greater than the forces of increase, the population would continue to decline and could become extinct.

Economic overfishing occurs whenever too much fishing pressure causes the potential economic benefits to be less than they could be. Many fisheries simply have more boats than needed to catch potential yield, and seasons have become shorter and shorter as more boats enter the fishery and catch the allowable harvest more rapidly. Far more money than is needed to catch the fish is spent on boat repairs, maintenance, fuel, and insurance. For example, governments may have subsidized vessel construction and fuel expenses or large fleets may have developed rapidly when the fisheries first began.

Related to any form of fishing is the ecological or ecosystem impact. Yet in that context there is no "optimal" level because, obviously, the actual number of fish in an ecosystem will decline continuously with increased fishing pressure; thus any amount of fishing can be said to be "ecosystem" overfishing, and to achieve the least possible impact means no fishing whatsoever. In some cases the total number of fish may be higher in a fished ecosystem if we remove important predators. However, any fishing is ecosystem overfishing to those with a focus on natural ecosystems.

But since we need to eat, let's look at abundance.

There is a relationship between the abundance of fish in an ecosystem and fishing pressure, sustainable yield, profit, and ecosystem impacts. When there is little or no fishing, there is little sustainable yield and precious little profit. As fishing pressure keeps increasing, first the profit peaks and then at higher fishing pressure the sustainable yield peaks. As fishing pressure further increases, both profits and sustainable yield decline. And when that happens we are said to be in a state of biological or economic overfishing. Normally we would expect profits to be highest when the fishery takes less than the biological yield.

What is a sustainable harvest?

"Sustainable development is development that meets the needs of the present without compromising the ability of future generations to meet their own needs," as defined by the Bruntland commission on sustainable development in 1987.

We speak of sustainable harvest as being able to continuously harvest a population or ecosystem in such a way that it can be maintained in the foreseeable future. We harvest a certain fraction of the population or ecosystem, and this fraction is low enough to allow the natural processes of birth and growth to replace what we take, on average, in the long term.

Problems arise when we think of a sustainable harvest as a constant quantity. This is almost impossible, as populations fluctuate naturally and harvests need to rise and fall with them. There are those who embrace extreme definitions of sustainability and argue that since petroleum resources are finite, no fishery that uses petroleum can be sustainable. We won't deal with that issue in this book.

To quantify *maximum sustainable yield* (MSY) we estimate the average of the catch that would be obtained when the stock is harvested at a rate that would maximize that average catch.

In some ways it is easier to think about what is not sustainable. Continuously taking more fish than can be replaced by reproduction and growth cannot be sustainable since the population will continue to decline until extinction. Any form of fishing that changes the ecosystem so that its underlying productivity is greatly reduced is not biologically sustainable. On the other hand, fisheries that require continuous subsidies to maintain profits are not economically sustainable.

Can fisheries be sustainably harvested?

The best scientific evidence shows that almost all fish populations can be sustainably harvested if the fraction taken each year is low enough and the method of harvest does not destroy the productive potential of the species or ecosystem. Many fish stocks were sustainably harvested for thousands of years mostly because social and cultural mechanisms kept the fraction harvested at a sustainable level or because technology did not yet allow fishermen to harvest too much. In the 20th century, and particularly in the second half, a number of changes took place. Advancing technology allowed boats to move farther from shore and fishermen to find the last refuges of many species. Modern communications, movement of peoples, and changing expectations often caused the breakdown of long-standing community-based management.

Is overfishing a new problem?

Overfishing has been with us since man first started fishing. Even with pre-industrial technology, natural resources could be overexploited, and we know that when humans first arrived in new parts of the world some of the more easily captured species were hunted to extinction. The historical record for fish is not as reliable as it is for land animals, but it is safe to assume that the most vulnerable species bore the brunt of first contact.

The concept of overfishing was already widely discussed in scientific circles in the second half of the 19th century. The British scientist Sir Norman Lockyer used the word in the journal *Nature* in 1877: "Nor does it seem to me quite worthy of my friend, in discussing the probabilities of *overfishing* in the sea, to try to prove his case by bringing forward an instance of *overfishing* in the rivers leading to a smaller supply of food." That overfishing involves taking too large a portion of a population was well understood by 1900, when Walter Garstang of Oxford University wrote, "We have, accordingly, so far as I can see, to face the established fact that the bottom fisheries are not only exhaustible, but in rapid and continuous process of exhaustion; that the rate at which sea fishes multiply and grow, even in favorable seasons, is exceeded by the rate of capture."

The biology of overfishing is always a question of the "rate at which sea fishes multiply and grow" compared to their "rate of capture."

As fishing technology got better, our ability to catch fish did, too, but the ability of the fish to multiply and grow stayed the same. Steam- and then oil-powered fishing vessels were the most important technological innovations. Trawl nets, which are dragged through the sea and were small when fishing boats still had sails, got ever larger as the fishing fleets switched to boats with ever more powerful engines after World War II. Other technological advances were made in fishing nets, especially cheap monofilament gill nets that almost anyone could afford. They are made of a near invisible mesh that traps fish behind their gills when they swim into the net. As these nets cost just a few dollars, their use spread around the world. Electronics such as global positioning systems (GPS) and fish-finders allowed fishermen to repeatedly find the same best fishing spots associated with reefs and rocks on the bottom and to do so in the fog.

We now have the technology to overfish almost every imaginable marine resource. The question is, do we have the political will and the social and cultural institutions to restrain ourselves?

Why does sustainable fishing reduce the number of fish in the ocean?

In the 1930s, Georgii Gause, a Russian biologist, did some very simple laboratory experiments to understand what limits growth of populations. He used the microscopic animal paramecium, which reproduces by splitting itself in two. Putting the paramecium into test tubes with plenty of food, he counted them as time passed and they increased. At first there was rapid splitting, but when the now numerous paramecia no longer had enough food, they slowed down and eventually stopped splitting and growing. Eventually the population reached an equilibrium called the *carrying capacity,* which is defined as the point when the number of splits and number of deaths (yes, every half of a half of a half of a paramecium eventually dies) were equal.

Wild populations are no different. Wildebeest in the Serengeti had once been decimated by rinderpest, a cattle disease similar to human smallpox. In the 1950s rinderpest was eliminated through vaccination and the wildebeest population began to grow again from a low of about 250,000 until it leveled off at 1,500,000 in the 1980s. Much like Gause's paramecia, once the population had increased enough in the 1960s and 1970s and the individual wildebeest had less to eat, the birth rate declined and the death rate increased until births and deaths were roughly equal.

Fishing clearly increases the death rate and, if nothing else changes, the population will go extinct. But as soon as there are fewer fish, the ones left in the sea have more food and other resources. Whatever limited an individual fish's growth before, be it food or good habitat with protection from predators, can be used to advantage once populations decrease. Eventually, deaths from predators may go down and birth rates may go up or a combination of both may come to pass. There is always a range of sustainable harvest rates that allows a population time to increase its birth rates and gives more fish a chance to live longer. Exceed that harvest rate too much and you are overfishing. For most marine fishes, the abundance that produces the maximum sustainable yield lies between 20% and 50% of the unfished abundance.

As long as we eat fish, there will always be fewer fish in the ocean than if we did not.

What is a collapsed fishery?

The word "collapsed" is most commonly used when a stock is at very low abundance measured either by some historical benchmark or by a theoretical calculation of how big the population would be without any harvesting. A population that is at 10% of the unfished abundance is generally considered collapsed.

When a population is at low abundance and at the same time birth and death rates change to make it impossible to rebuild that population even though harvesting has stopped, we are looking at a much more complex form of collapse.

What happened to the Canadian cod?

On July 2, 1992, the Canadian Minister of Fisheries and Oceans John Crosbie announced that the cod fishery of Newfoundland would be closed. A legendary 500-year-old fishery of seemingly everlasting bounty and the economic backbone of the

Province of Newfoundland and Labrador was finished. Since then it has been the icon of the crisis facing the world's fisheries.

Even before Columbus came to America, Basque fishermen sailed to the Grand Banks to fish for cod. Cod was the reason for the settlement of Newfoundland, where fish means cod. After being prudently harvested for 500 years, that vast biomass of fish was reduced from millions of tons to a small remnant of tens of thousands of tons in three decades. Only recently, eighteen years later, have there been signs of rebuilding.

The cod collapse caused undreamt-of social upheaval. It suddenly put 20,000 people out of work. The Newfoundland economy nosedived, Canadian taxpayers paid over $1 billion CAD per year in support payments to offset the loss, and an island culture built on cod was deeply shaken.

Why did the Canadian cod collapse?

We caught too many. For hundreds of years the Newfoundland cod population easily sustained catches between 100–200 thousand tons because births and growth were reasonably balanced by natural deaths, death by predators, and fishing. Very likely less than 10% were removed each year, a level that the population could support. But as soon as large foreign factory ships dramatically increased the catch, in the 1960s, to well over 30%, reportedly taking as much as 800,000 tons in one single year, births and growth could no longer keep up and the population declined.

By the time Canada took control of the fishery in 1977, the total weight of mature spawning fish was down to a few hundred thousand tons, from over 1.5 million tons in 1962. Canada lowered the total catches, and initially the population increased. But through the mid- to late-1980s the population stopped increasing and apparently stabilized at about 25% of what it had been in 1960. But then, in the late 1980s, the number of young fish suddenly declined, and individual fish grew more slowly and died faster. The target catch for 1991 could not be taken because it was greater than the entire population.

Competing explanations for the ultimate collapse are plentiful. Many believe it is a simple case of overfishing; the population was finally small enough that not enough eggs were being produced, not enough young being born. Others blamed it on too many seals. Then again, there is substantial anecdotal evidence that the catch was much higher than reported because smaller, less valuable fish were thrown overboard (*discards*), and thus not counted. Finally there was, at the same time, a decline in ocean temperatures and a change in the tiny zooplankton that are the base of the food chain that did not favor the cod and that may have contributed to the declines in births and growth rates and the increases in mortality.

It was a terrible time to be a cod in Canada in the late 1980s and early 1990s. For management purposes, individual populations of Eastern Canada cod have been aggregated into about a half dozen groups called "stocks." In the late 1980s, even stocks that were plentiful and growing throughout the 1970s and early 1980s suddenly stopped increasing and the net difference between births, body growth, and natural mortality went negative. These populations would have declined without fishing. From what we now know there is nothing managers could have done to stop this falling off in productivity of the cod, but what they failed to do is cut back the catches in time to prevent all Canadian cod populations from being driven to very, very low levels.

Are all cod fisheries collapsed?

In the 1990s almost all the world's cod populations were overfished and pushed to very low abundance. Many, if not most, would have been below the 10% level. Almost all the European stocks, although driven to equally low abundance, remained very productive and often sustained 30%–50% annual exploitation rates with little or no continued decline. These stocks rebuilt once fishing pressure was reduced.

Not so the Canadian stocks. Despite very low harvests they did not rebuild.

The two American cod stocks were also overfished to low abundance but continued to be productive. Indeed, both the Gulf of Maine and the Georges Bank stocks are rebuilding but remain below target levels.

Two cod stocks never collapsed. The Barents Sea stock, located in the north of Norway and shared with Russia, is the largest cod stock in the world. It was estimated to be over 4 million tons in 2010 and is not overfished in any sense. The Icelandic cod stock is below the target levels of abundance but has never collapsed and is currently fished at what is thought to be the rate that would produce maximum sustainable yield.

2
HISTORICAL OVERFISHING

Is overfishing a new problem?

Although they are mammals, whales were originally thought to be fish and their exploitation was referred to as the whale fishery. As governments replaced kingdoms and fisheries regulations came into being, management of whale fisheries became the province of fisheries departments. Known for its overexploitation, the whale fishery provides an excellent example of how overfishing proceeds and some of the inevitable consequences that surround it.

The thousand year history of commercial exploitation of whales by Europeans illustrates many aspects of overfishing. Basque whalers were active by the 12th century, and their primary target was the northern right whale—"right" because it moves slowly and does not sink when killed. Initially Basque whalers stayed along the coast and harvested meat from stranded whales. The next step was to spot whales from the shore, put out small boats, pursue the whales, and strike them with harpoons much like the common images of 19th-century whalers seen in various versions of *Moby Dick*. As sailing technology improved and local concentrations in the Bay of Biscay had been depleted, these earliest European whalers moved into bigger boats and traveled farther north. By the 17th century they reached the high Arctic around

Spitsbergen, where they killed both right whales and a near relative, the bowhead whale. News of the wealth from these Arctic voyages spread, and soon the English, Dutch, and Spanish were regularly sending their vessels north.

In addition to advances in sailing and navigation that enabled longer journeys, there was a shift from processing blubber into oil at shore-based facilities to processing on board the ships. This freed the ships from the ties of land and, in the 17th century, made open-ocean whaling possible. At the same time around almost all the world, whaling continued to be a locally important economic activity along the coasts. In New England the early Yankee whalers were much like the Basques of the 12th century; they spotted whales from shore, chased them in small boats and brought them back to shore for processing. In Japan that kind of whaling dates from at least the seventh century.

By 1690 the stocks of whales in the eastern Arctic were depleted and whaling fleets moved west to Greenland and then much farther afield. The discovery of a new population of bowhead whales in the Bering Sea in 1848 led to another gold rush, and these whales were rapidly depleted. By the mid-19th century almost all the world's oceans were being explored by whalers from many countries.

The late 19th century led to a number of important technological changes. The market for whale oil declined when petroleum began to fuel lamps. By the end of the 19th century the sperm whale fishery that had supplied most of the oil had almost completely collapsed due to lack of markets. Counteracting the decline of markets were major technological developments, especially the development of explosive harpoons. Fired by guns mounted on high-speed steam-powered vessels, these harpoons could both attach to and kill the larger whales. Explosive harpoons replaced the traditional techniques of attaching a drogue or small boat to a whale with a harpoon, letting the animal exhaust itself by pulling against it, then drawing alongside to kill it by repeatedly driving a lance into its body (the act that killed most whalers). Fin and humpbacks were especially difficult and dangerous to kill with a lance. By combining steam power and explosive harpoons, coastal whalers in New England were able to capture fin and humpback whales and drag them to shore, although many were lost because not all of them float. The Norwegians perfected modern industrial whaling, adding air inflation of carcasses to the explosive harpoon and high-speed steam-powered catcher boats. This technology allowed them to exploit the mother lode of whale populations, the blue, fin, sei, and humpback whales of the Antarctic. Beginning in the early 20th century, annual catches of great whales (these species plus the sperm whale) rose to over 70,000 per year, almost all from the southern ocean. Catches of the most valuable, the blue whale, peaked about 1930. Attention then shifted to the less valuable species. The catch of fin whales peaked in the 1950s, sei whales peaked in the 1960s, and the small minke whale—weighing less than one-tenth of the great blue whales—became the mainstay of Antarctic commercial whaling in the 1970s. By the end of industrial whaling, the larger right, humpback, and blue whales were almost extinct and most other large whales were heavily depleted.

There is an important lesson to be learned here. In an unregulated fishery, fishing will continue as long as it is profitable and will ultimately be limited either when the target species is so rare that the returns no longer pay the costs or by the market's decline.

The history of whaling illustrates a phenomenon seen in many of the world's fisheries known as *sequential depletion*. Fishing starts close to home ports and on the most vulnerable species. As the initial target species or areas are depleted fishing moves farther afield. Fisheries of new populations or species are developed in response to local declines.

Whaling also highlights the problems of international management of the high seas. In the Western tradition anyone is free to fish on the open ocean, a doctrine known as *Freedom of the Seas*. However, countries do recognize that they must cooperate and try to prevent overfishing. The *International Whaling Commission* (IWC) was established in 1946 in part to respond to concerns that whale oil markets were being flooded with oil from Antarctic whaling. In the 1960s an expert panel of international specialists, known informally as the "three wise men," was brought in to advise on the sustainability of harvest levels. They recommended immediate reductions in harvest. However, the IWC proved unable to effectively regulate the catch. Some individual countries did not abide by the agreed-on regulations and there was no independent enforcement. As commercial whaling was reduced in the 1970s and 1980s, various large whale stocks began to recover, and in 1985/1986 the IWC implemented a zero quota on all commercial whaling, usually referred to as a moratorium.

Can whales be sustainably harvested?

Sustainable harvesting requires that the target population have the potential to increase in abundance. The sustainable harvest is the amount you could remove each year and keep the population at the same level. For instance, the gray whales off California increased from the late 1960s to the 1990s from 12,000 to about 20,000, a rate of about 3%–4% per year. It is now thought that the gray whales in the eastern North Pacific may have rebuilt to the abundance that existed before Western industrial whaling (they are extinct in the North Atlantic and close to extinction in the Western North Pacific), while at the same time there is an annual Russian subsistence hunt of about 120 gray whales per year.

If, for example, there had been a harvest of perhaps 2% per year, the population would have grown more slowly and a few hundred additional animals a year could have been taken while the population was rebuilding. Of course, this assumes that the population estimates were roughly correct and there was a method in place to assure that the actual catch did not exceed 2% per year. The reason for much of the concern with resumption of commercial whaling is the history of poor compliance with international agreements, not only for whales, but in many international fisheries.

For quite a long time now the scientific committee of the IWC has attempted to develop rules that would provide the scientific underpinning of sustainable whaling in case the Commission decides to end the moratorium. In particular, they have sought to develop a "harvest strategy" that would take into consideration all the uncertainties about whale populations. What is their rate of increase? How many whales are really there? At what age do they breed and how well do their young survive? How are their populations structured? We know little about the boundaries between breeding populations—there may be separate sub-populations that must not be depleted by harvests.

For many years teams of scientists have evaluated the potential of different harvest rules to meet the conservation objectives. In the early 1990s, the Scientific Committee of the IWC and then the actual Commission adopted the scientific elements of a harvest control rule known as the *Revised Management Procedure* as an approach to sustainably harvest whales. However, the Commission also said that before any potential harvest, additional agreements about monitoring and compliance need to be adopted. This has not happened. And while the moratorium is in place, the Revised Management Procedure has not been implemented under the auspices of the IWC, although a slight variant is used for the Alaskan bowhead whale harvest.

Norway does legally hunt minke whales commercially in its own waters and currently takes about 1,000 minke whales a year out of a population estimated to be well over 100,000. This is believed to be biologically sustainable. Even though the *International Convention on Trade in Endangered Species*

(CITES) has banned international trade of minke whales, there is a technicality in the treaty that allows Norway to export minke whale meat legally. Iceland hunts whales under the same objection procedure, whereas Japan's whaling falls under provisions for scientific research. The United States is also a whaling nation. Each year about 50 bowhead whales are taken by native Eskimos on the North Slope and western coast of Alaska out of a population estimated at over 11,000. Although the bowhead population is listed as "endangered" under the U.S. Endangered Species Act, the numbers are growing and the level of take and the status of the population is routinely reviewed by the IWC. There is no commercial trade in these whales other than the sale of native handicrafts, and all the edible products are consumed locally.

How do we estimate the abundance of animals in the ocean?

John Shepherd, a well-known marine scientist, once quipped that "counting fish is just as easy as counting trees, except they are invisible and they move." Counting mobile marine animals like whales and fish is very hard and a great deal of scientific energy goes into trying to do so. In general we rely on a broad range of survey methods that provide a relative index of abundance, rather than an absolute estimate. The most common method used in studies of fish populations is to design a scientific survey that samples the different habitats in a systematic fashion, and then use some method of detecting the fish to provide the index. Most common for bottom fish is to use trawl nets. Other methods include using sonar to detect the fish or cameras to photograph them. Another method is to tag thousands of individual fish and when they are later captured to see what fraction of those are tagged. Sedentary animals like abalone, clams, and scallops can often be more reliably estimated by systematic sampling of the sea bottom.

Two techniques are commonly used for whales. Photographs of individual whales can be taken and individuals can be recognized by distinctive marks. For some commonly seen populations, almost all animals may be photographed each year. Ships sailing on pre-specified lines called "transects" will count how many whales are seen per kilometer traveled. This provides both a relative index of abundance, and, using various calibration techniques, absolute densities can be estimated.

Often scientists have several of these techniques available; they may have surveys, tagging data, and the age distribution of the population. All of these data are often combined in a *stock assessment*, a statistical procedure that estimates historical trends in abundance of the population. These stock assessments provide the basis for regulations set by most fisheries management agencies.

Can scientists estimate the sustainable yield?

A stock assessment will show the number of individuals of a population over time. Essentially it is an accounting framework for tracking births, deaths, and the growth of individuals. Scientists often calculate *surplus production*, the net increase in the biomass of the population from year to year, plus the catch. If the number of animals in a harvested population is stable, then the surplus production is simply the catch. If a population increases, the surplus production is the amount of increase plus the catch. The sustainable yield at any population size is the average surplus production at that population size. What emerges from a stock assessment is the history of abundance and surplus production. The amount of allowable harvest will normally be related to the surplus production. If the stock is believed to be at the target abundance, the recommended catch will be the estimate of the surplus production at that population level. If the stock is thought to be depleted, then the recommended catch will be less than the surplus production to allow the stock to rebuild toward its target level.

Is there any value in Japanese "research whaling"?

The government of Japan authorizes killing certain whale species from certain populations as part of a scientific research program and the meat is then sold in Japan. Scientists on board collect biological data such as size, gender, age, pregnancy status, and food habits. Tissue samples are collected to study population structure and exposure to certain contaminants. Survey vessels collect abundance data by recording how often whales are seen.

Critics argue that the scientific output of this research effort measured in peer-reviewed publications is very low, and with the Revised Management Procedure, data collected from dead whales are not needed to manage commercial whaling should the moratorium be dropped. The Japanese government argues that CITES allows research whaling when authorized by a member government that also has the authority to set the limits of its whaling. Moreover, Japan claims that research is needed to reduce uncertainty about productivity, competition between species, and the impact of pollution on whales in the Southern Ocean and western North Pacific.

Japanese research whaling is a contentious and often emotional issue. Some see it as thinly veiled commercial whaling, others as a means to provide valuable scientific data.

Certainly it does not take enough animals to threaten the populations. The major conservation concerns come either from an animal rights view—killing any whales is bad—or from concerns that it is the "thin edge of the wedge" that may lead to a resumption of commercial whaling. The most serious charge, however, is that many scientists regard scientific whaling as a pretense that allows Japan to flout the zero-whaling agreement.

Is depleting one population and moving on to the next a common problem?

As industrial fishing expanded, fishing fleets would deplete resources in one fishery and then search for and find the next. To a great extent those days are over, because there really are no new significant resources to exploit and most effort now goes into trying to sustainably manage the marine resources that are left. While new fisheries are always being developed and boats are going deeper and farther afield, all newly discovered fish stocks have been very small and none have contributed significantly to the world catch in almost 20 years. The bulk of the world's fish catch in 2010 came from the same species and stocks it did in 1990.

There is a related and ongoing problem due to the reduction of fishing fleets in Western countries. Governments often pay boats to leave their fisheries, with a restriction that the same boats cannot be converted to fishing boats in another fishery in the same country. The result is that these boats often start fishing elsewhere, possibly illegally in the waters of other countries that have less restrictive management systems.

3

RECOVERY OF FISHERIES

Can fish stocks recover from overfishing?

Meet the striped bass. Abundant, tasty, and fights like a tiger. The perfect sport fish, Methuselahs of 70 pounds are still being caught and stand as the poster boys for a wondrous recovery from overfishing.

When Europeans first arrived, the abundance of striped bass seemed limitless, like so many other resources in North America. Captain John Smith felt he could walk "dry shod" on their backs across the river—though that may have been just a bit exaggerated in order to attract more colonists.

In 1639, the Massachusetts colony was already concerned enough to forbid the use of striped bass as fertilizer for crops. Throughout the 18th and 19th centuries as well as much of the 20th both sport and commercial fisheries of striped bass were extremely important. The Chesapeake Bay and Long Island Sound fisheries were particularly large and there were smaller fisheries down the east coast all the way to Florida.

There were periodic declines in abundance. In 1905 striped bass were considered "uncommon" around Woods Hole, Massachusetts. Once reliable data were collected, it became clear that catches peaked in 1973 followed by a 90% decline in abundance, catch, and catch rate for both commercial and recreational fisheries. One of the most important fisheries on the east coast had collapsed.

Beginning in the mid-1980s severe catch regulations were imposed. They ranged from a complete ban on striped bass fishing in Maryland and Delaware to big increases in the minimum size limits as well as reduced daily catch limits coastwide. These regulations, combined with ongoing freshwater habitat improvement, led to a remarkable recovery. By the mid-1990s the abundance of spawning females had increased 10-fold and juvenile abundance was at record levels. The Chesapeake stock was declared rebuilt in 1995, as was the Delaware River stock in 1998. The fishery is now more valuable than ever.

Striped bass are *anadromous*; they spawn in fresh water and their eggs drift downstream or with the tides after they hatch. Some juveniles move farther downstream into estuarine areas where they grow for two to three years. Finally they move out into the ocean where they stay for their adult lives. The females return to spawn in freshwater after they have reached sexual maturity at four to eight years of age.

Stripers continue to grow as long as they live. The record for the largest one is 125 pounds. To complete their life cycle they return to their traditional spawning sites, and to spawn successfully they need good habitat where the juveniles can find refuge from predators. The Chesapeake Bay with the rivers that feed into it is still the most important striped bass habitat and produces 75% of the east coast stock followed by the Delaware and the Hudson rivers.

The decline of striped bass was a classic case of "death by a thousand cuts." Freshwater habitat has been degraded and lost since colonial times. During the industrial revolution, weirs diverted water to mills, and the 20th century brought intensive pollution from large industries, agricultural runoff, and urban sewage. Worse yet, acid rain drifted in from the big industrial areas of the Midwest. Commercial and sport fishing increased further and took fish that were so small they had not even spawned yet. Size limits were set at 12–14 inches, but females need to grow to at least 24–28 inches to mature. Fishing was so intense that over half of all fish were caught each year. To add even more insults, large power plants on the Hudson and Delaware killed eggs and juveniles where the river water passed through their cooling systems.

When a species is dying from a thousand cuts, a lot of bandages are needed. And the many bandages kept the patient alive. Water quality improved dramatically after passage of the Clean Water Act in the 1970s, and as soon as fishing pressure, the worst offender, was severely reduced, the stripers quickly bounced back and grew to spawning age.

There is also a clear influence of climate on recruitment success. In the 1970s, when striped bass collapsed, weather patterns were dominated by relatively warm and dry winter-spring conditions. When weather patterns shifted to colder and wetter winters and springs, the striper population recovered.

The key to the recovery was coordination of effort. Because striped bass spawn in many states and are caught in both state and federal waters, no single management authority had control. Habitat improvements in Virginia would be pointless if New York or Maryland were to catch too many fish. A series of federal laws and inter-state agreements made the needed coordination for the rebuilding plan possible. Even though such substantial coordination was difficult, there was the unquestioned need for change. Everyone recognized just how close to extreme danger the populations were and that rebuilding was absolutely necessary.

In sum, overfishing and bad weather were responsible for the recruitment failure, and both good management and better weather must be given credit for the recovery.

However—and there is always a however somewhere to keep us from getting too smug—even the best stories in fisheries management have no happy-forever-after ending, and so it is with striped bass. It is true that the stripers have recovered remarkably well, yet they are now threatened with a disease, mycobacteriosis, that infects half or more of the mature stock and may eventually be fatal. In recent years the weather again seems to have turned against them and populations are declining once again.

How important is habitat to fish populations?

Lose the habitat, lose the fish.

We can stop overfishing, but without hospitable habitat, fish populations will not rebuild. In the United States, the Magnusson-Stevens Fisheries Management and Conservation Act specifies that "Essential Fish Habitat" must be protected. But how to define "essential?" Fish need water with just the right physical conditions such as temperature, salinity, freedom from toxic chemicals, and pH, which is a measure of the level of acidity. pH has been a big concern in lakes and streams that are in the path of acid rain. Lately even the oceans are becoming more acidic due to atmospheric CO_2. Many fish are fussy about where they lay their eggs and want their freshwater bottom to be just so, and, once hatched, the juveniles must have the proper food to grow well, and places to hide from predators.

Habitat changes come in many shapes and sizes. Dams that block the passage of salmon, shad, and other anadromous fishes to their spawning habitats obviously create a total loss of habitat. Acid rain, increasing temperatures, and low levels of pollution may only reduce the quality of the habitat and lower the overall survival rate of fish. In general, the many years of building dams, polluting rivers, and diverting water to thirsty cities and fields have left a heavy imprint on freshwater fish habitat. We have managed to transform estuaries beyond any recognition with dikes, coastal development, expansion of cities, and pollution. On the whole, the more people, the greater the habitat degradation. But, then again, we have made it possible for even far away events such as the

Exxon Valdez and Deepwater Horizon oil spills to have large
and lasting impacts on habitats in remote areas and far
offshore.

What about the enormous numbers seen by John Smith?

Early accounts don't give precise numbers, but they do give
us some scale of what we have lost. There may have been
some exaggeration—sweeping statements do attract attention—but there is no doubt that many fish populations have
dwindled where habitats were degraded or lost or where they
were overfished.

Daniel Pauly, a French scientist now working in Canada,
introduced the concept of *changing baseline*. Each generation
thinks of "natural" conditions as the good old days, how
things were when they were young. We must be careful to
establish a true historical baseline of abundance and not one
of our greener days, 20 or 40 years ago. To this end scientists
and historians are already working on establishing historical
abundances using the tools of paleontological and historical
research.

What is the difference between recruitment overfishing and growth overfishing?

Recruitment and growth overfishing are the two components
of *yield overfishing*.

The number of newly spawned fish that survive at least for
the first year is known as the *recruitment*. Once there are no
longer enough spawning fish to produce enough recruitment,
we talk about *recruitment overfishing*.

According to ecological theory, recruitment is ultimately
limited by habitat conditions such as available food and
refuge from predators. At some point there may no longer be
any room at the inn and some eggs and larvae will run out of
habitat. With light fishing there may still be enough spawners

left for their eggs and larvae to use all the available habitat. With heavier fishing, suddenly recruitment is no longer limited by the habitat but by the number of eggs and larvae that survive. How many of those do survive depends of course on how many mature fish were able to spawn, and once there are too few of those we enter the realm of *recruitment overfishing*.

Growth overfishing has to do with how old a fish is when captured relative to its size.

Fish typically grow very fast when they are young and then their growth slows down. When they start to mature and produce eggs and sperm, they use more of their energy for reproduction and less for growth. If we had complete control over the age at which we catch fish, we would catch them only after they stop growing. Theoretically the best time to catch them is when their rate of growth is exactly the same as their probability of dying from natural causes. When we catch small fish that are still growing fast we are *growth overfishing*. We are wasting the potential growth of each small fish we catch too young. Since most fishing gear does not allow us to catch fish of a specific size or age, how much growth overfishing there is essentially depends on how hard we fish.

How much we overfish the yield depends on a mixture of recruitment overfishing and growth overfishing. We have learned through both theory and empirical studies that there is a level of fishing pressure that gives us the best long-term sustainable yield. This depends, of course, on assumptions of a stable environment, and we are only too aware that both man and nature itself are constantly changing the world around us. (For discussion of the impact of climate, see chapter 6.)

Can recreational and commercial fisheries co-exist?

In 2007, striped bass were declared a game fish in federal waters after intense political lobbying by the recreational fishing industry. This declaration, prohibiting all commercial fishing in the ocean beyond state marine boundaries, was somewhat meaningless since no fishing for striped bass had been allowed in any federal waters to begin with. The situation was only too symptomatic of the conflict between recreational and commercial fishermen. Meanwhile, lobbying continues in pursuit of further declarations of ever more highly prized recreational fish as game fish and thus off limits for commercial fishing.

Recreational and commercial fisheries can co-exist, but co-existence is often oh so frustratingly difficult. Nothing infuriates recreational fishermen more than seeing a commercial fishing boat hauling in hundreds of "their" fish. Commercial fishermen become nearly apoplectic when they hear recreational industry representatives firmly declare that their people take only one or two fish each and all conservation problems are caused by commercial fisheries.

For conservation purposes it is irrelevant who kills the fish—a dead fish is a dead fish.

In developed countries, recreational fishermen generally catch few fish relative to the size of the commercial catch. But for the most prized fish the picture is drastically different. These fish are also often the most heavily fished species, and recreational fishermen commonly take half or more of the total catch.

But this book is about overfishing, not the allocation of fish to different user groups. I will close this discussion by saying that because of the overwhelming number of recreational fishermen and the political power of their lobby they tend to win the battles over who gets the fish.

4

MODERN INDUSTRIAL FISHERIES MANAGEMENT

What is an example of a well-managed fishery?

On September 10, 2009, the *Economist* published an article on Bluefin tuna and Eastern Bering Sea pollock entitled "A Tale of Two Fisheries: How to Pillage the Oceans Deliberately, and by Accident." The article said: "There are two ways to overfish the sea. One is to ignore scientific advice and plunder on regardless [*referring to bluefin tuna*]. The other is to accept the advice, and then discover it isn't good enough [*referring to Eastern Bering Sea pollock*]." The Eastern Bering Sea pollock fishery, the pride of the U.S. management system, was held up in the international media as badly managed. Greenpeace, the major source cited by the *Economist,* had been arguing for a year that the fishery was on the verge of collapse. Their website in October of 2008 said, "Just as the financial institutions on Wall Street collapsed due to poor oversight and mismanagement, the pollock fishery is on the fast-track to collapse as well....Each year, fishery managers want to catch large amounts of fish to get to the most profit. They fish and fish, even when science tells them the fish can't keep up and they need to cut back."

The cause of concern was the decline in the abundance of pollock to 4.1 million tons in 2008 from a high of 12.8 million tons in 1995. Catches had been reduced from 1.5 million tons to 0.8 million tons. Environmental groups interpret that trend as a collapse under way—the Northern cod story repeating itself. The National Marine Fisheries Service, the U.S. government agency in charge of science for Eastern Bering Sea pollock, argued that the decline in abundance was natural. The 12.8 million tons were a historical high, reached after several years of particularly good survival of young fish. They argued it was unlikely the stock would stay that high for long, especially given signs of well below average conditions for young fish (due to several years of poor conditions for survival through early life stages). The scientists' expectation was that conditions would fall below average and, given evidence of recent better conditions, the stock would then rebuild. The scientists were proved right: by 2011 the stock had rebuilt to 9.6 million tons, and the recommended allowable catch was increased to nearly 1.3 million tons.

The Eastern Bering Sea pollock fishery is a large-scale industrial fishery with the catch split 40% to a fleet of 16 large factory freezer trawlers that process all their fish on board and 60% to a fleet of 82 *catcher boats* that deliver their catch to shore-based plants or to three *motherships* that process at sea. The available pollock data are the envy of most fisheries managers because in order to manage a fishery well you want to know trends in abundance from scientifically designed surveys. For pollock, there are two surveys a year. A manager needs to know how many fish are caught and, if any, how many are discarded (or thrown overboard). In this fishery there are two observers of catch and discards on board each large vessel, which means that fishing operations are watched around the clock. There is close to 80% observer coverage of the catcher boat fleet. Moreover, there is also a large-scale research program on the status of the ecosystem of the Eastern Bering Sea that studies species that are not commercially important, with the goal of understanding the big picture of the eco-system. Ever since it became an American fishery through the 200-mile economic zone, there has been a total catch cap of

2 million tons for all species from the Eastern Bering Sea and Aleutian Islands. This has meant that in most years the actual allowed catch of each species has been less than what scientists said was biologically acceptable. For instance, in 1991 the recommended allowable biological catch was 1.7 million tons, but the actual catch was only 1.2 million tons. Finally, unlike the fisheries of the north Atlantic, the pollock fishery is new, and the data from it go back almost to its beginning in 1977. Whereas the Atlantic fishery managers are always uncertain about the abundance of their species in historical times, in Alaska we know that pollock reached their highest abundance ever (or at least since fishing began) in 1995.

What is different about the pollock fishery that makes it such a good example of sustainable management?

The pollock fishery stands out because the data are excellent, the harvest control rules are conservative, and there is an ecosystem-wide cap on total catch. Few fisheries in the world have this level of observer coverage and frequency of surveys. The harvest control rule allows for a relatively small fraction of the total stock to be harvested. The average since 1991 has been 15%. Finally, the ecosystem-wide maximum catch of 2 million tons provides some security that the entire ecosystem will not be nearly as heavily impacted as is common elsewhere.

Why does the allowable catch change so much from year to year?

Once scientists determine the pollock's abundance, the allowable catch is calculated from a published *harvest control rule*, which is one part of the Fisheries Management Plan (FMP). This rule is quite simple: If the abundance is above a certain target of biomass, the allowable catch is a fixed percentage of the stock biomass. This percentage is thought to produce maximum long-term yield. If, however, the stock drops below the target level, then the harvest is reduced and will be zero if the stock reaches a minimum *limit* biomass. At that point all fishing will be stopped. This rule was designed to provide long-term maximum yield when conditions are good and scale back fishing pressure when times are bad.

Even without fishing, fish stocks vary considerably from year to year. There are good years when environmental conditions are right and fish grow and survive well, and there are bad years of poor growth and survival. These good and bad years often come together instead of being randomly dispersed. For pollock the 1990s were particularly good years but the early 2000s were poor. Consequently, abundance rose in the 1990s and fell in the 2000s and so did the catch as the natural consequence of a sustainable fishery management plan. The declines in catch seen in pollock are exactly what should happen in a good management system and are decidedly not symptoms of a stock collapsing.

There are of course those who try to use catch as a measure of a fish population's health. Now, if the catch were always the same percentage, that would be correct, but when this percentage is changed at low abundance, declines in catch will be much steeper than declines in abundance. However, the 2 million ton cap means that at high abundance harvest rates on pollock are very low and one actually sees an increase in harvest rates when the stock declines from very high abundance.

What is a stock assessment?

Stock assessment is a scientific process in which all available data on a fish stock are combined to estimate what the historical trends in abundance are, what the percentage harvested is, and how productive this stock has been. The data include everything that is known about how abundant the fish have been in the past, how big or small the catches were, what size the fish were as well as their average age and length, called age and length distribution. At the core of the analysis is usually some form of mathematical calculations that uses these data to estimate the number of births and deaths. Generally, a small team of scientists does the initial calculations, followed by several levels of review. In the case of Alaska pollock, an established analysis has been used for several years. Every year, all the new data on catches, surveys, and age distribution are used to update the status of the stock. This analysis is then reviewed first by a *plan team* that consists mostly of state and federal scientists, and then again by the *Scientific and Statistical Committee* of the North Pacific Fishery Management Council. This committee is made up of a number of independent academic participants and other government scientists and provides a further level of peer review.

What is an observer program?

An observer's job on a fishing boat is to record data about the fishing operations. Typically an observer records the time and place of all fishing activities, the catch brought on board, as well as anything tossed over the side. Catch of species not meant to be fished such as marine birds and mammals are of particular interest because of ecosystem concerns. Often observers are responsible for taking scientific samples of the catch, which usually involves determining the number of fish caught by species and measuring the lengths of a sample of the fish. They may also take samples of the fishes' ear bones, the "otoliths," which have annual growth rings much like trees and are used to determine the age of the fish. In some systems, observers have a scientific role only, whereas in others they are responsible for enforcement of regulations.

Why are there not more observer programs in world fisheries?

While observer coverage is very high in the pollock fishery, there are many fisheries without any observers whatsoever.

Even in large industrial fisheries of countries like New Zealand the observer coverage may be less than 10%. To really understand a fishery there must be observers, particularly to record what is thrown overboard. The landings, or how many fish are kept, can be determined by sampling boats when they come into port and the locations fished can be determined by satellite tracking systems. But any reliance on fishermen's reports to understand what is discarded would be foolish, as there are usually strong incentives to fudge those numbers.

Why, then, are there not more observers in fisheries? First, they are expensive. Second, fishermen generally do not like having observers on board, especially if they also have a role in enforcement. Third, it is often difficult to find space for an observer on a small boat. Finally, good observers are difficult to retain. The job is interesting but hard, and long periods at sea are spent away from family and friends.

Automated cameras, already used in a number of fisheries, promise to get around the problem of too few observers. Recording continuously, they cover the entire fishing deck and, depending on the nature of the fishing, they make it possible to see what is discarded, identify species, and even sometimes measure the lengths of individual fish. Obviously cameras are no substitute for biological sampling, but it seems likely that in the future many more fisheries will use them. And that is a step forward.

What is a certified fishery?

We are accustomed to the existence of, and often demand, an agency to certify that a given product meets a certain standard. We like our meat to be certified fit for our consumption, our children's toys to be safe. In fisheries, the most visible certification agency is the Marine Stewardship Council (MSC), a nongovernmental organization (NGO) originally founded by the World Wildlife Fund and the large food company Unilever. The MSC has established a set of standards that must be met for a fishery to be considered "well managed." Once a stock has been certified, its fish can be sold with the MSC label, indicating they have been fished sustainably. Many large food chains, including Walmart in the United States, have made commitments to sell only MSC certified seafood, and this in turn provides incentives for fisheries to meet the MSC standards and seek certification.

Certification is a complex and often controversial process. MSC certification works as follows: A group with a major stake in a fishery, usually a government agency or a fishing industry association, applies to the MSC for certification. As a first step, they employ one of several independent "certifiers" to manage the process. These certifiers are consulting companies that have been accredited by the MSC to do the managing. The certifier then employs a team of consultants (normally three) to score the fishery according to criteria the MSC has set. These criteria encompass the state of the stock and the scientific data available for its management, its management system, and the ecosystem impacts of the fishery. If the fishery gets a passing score, it is certified. There is a minimum score for each criterion, and if the fishery fails to meet any one or more, the certification may be conditional on meeting the required score within a specified period. Many certified fisheries will have had several dozen conditions attached to their initial certification.

Once the certification team has done its scoring, the evaluation is put out for comment by both the clients and any other interested stakeholders. The scoring may be redone based on their comments. This evaluation is further reviewed by a second team of independent scientists. Finally, if the client organization or a stakeholder is unhappy with the certification, it may appeal and the MSC sets up an "appeals panel" to evaluate the issues being appealed.

Once certified, there is an annual review of whether the conditions have been met and if there are any changes to the fishery that would change its scoring. The Alaska pollock fishery is the largest fishery in the world to be MSC certified (in 2005) and it was recertified in 2010.

MSC certification has been extremely controversial. The fishing industries think that the standards are too high and the costs are excessive. Environmental NGOs believe the standards are too low and the process is seriously flawed because the costs are often borne by the fishing industry itself. Certifications that have been particularly controversial are for Patagonian toothfish off South Georgia Island, discussed in chapter 12, and Antarctic krill. By 2011, 102 stocks of fish had been certified, constituting 12% of the world's catch of fish for human consumption, and another 142 stocks were in the process of certification.

Why do some NGOs believe the Eastern Bering Sea pollock fishery is not well managed?

Nongovernmental organizations such as Greenpeace and Oceana have three key concerns about the pollock fishery. In their eyes the decline in abundance and catch in the 2006–2009 period is a sign of poor management and means that overall exploitation rates were excessive. More important, they are concerned that the pollock fishery removes food for marine mammals and birds, particularly the Steller sea lion that is listed as an endangered species and underwent a substantial decline in the 1960s–1980s. The population in the Aleutian Islands seems to still decline slightly. Moreover, there have been long-term declines in many other marine birds and mammals in the Eastern Bering Sea and Aleutian Islands, and fishing for pollock and other fish is thought to have reduced their food supply.

Finally, the pollock fishery catches a number of other species, with special concern for salmon. This is an interesting and complex issue. The pollock fishery has one of the lowest rates of by-catch (tons of non-target species caught per ton of target species) in the world, but because it is such a high-volume fishery the total by-catch is considerable. Tens of thousands of Chinook salmon, for instance, have been caught in some years. Given the by-catch rate, the pollock fishery appears to be a good example of ways to produce food from the ocean with little impact on other species—but when counting the individuals of non-target species, the pollock fishery itself becomes a target for concerned NGOs.

5

ECONOMIC OVERFISHING

Is overfishing only a biological problem?

The Pacific halibut fishery has long been considered the outstanding success of sustainable management. The International Pacific Halibut Commission was formed in 1923 by the United States and Canada to jointly manage the halibut stock on the Pacific coast. This stock has been healthy and has not been considered overfished since the 1940s, and was at record abundances in the 1990s.

Yet not everyone was happy. The main fishery in Alaska was an *open access fishery*. Anyone wanting to fish could get a license for a nominal fee. Inevitably, the number of fishing boats rose from a few hundred in the 1950s to 4,000 in Alaska alone in the 1980s. But since the total catch could not be increased, the fishing season had to be shortened accordingly. It went down from four or five months in the 1960s to a single day in the early 1990s in several places. It became a "derby" fishery where thousands of boats dropped their gear at the sound of the opening gun and hauled it back up 24 hours later.

The fishery was also dangerous. If there happened to be a storm on the one open day, fishermen had to make a brutal choice: stay at home and forgo their income or go out and risk their lives. Many died who went fishing.

Halibut are caught on *longlines*, strong fishing lines, several miles long, with evenly spaced baited hooks, that lie on the ocean floor. Quite often boats put out more lines than they could retrieve before the closing gun. Those abandoned lines continued "ghost fishing" until the bait was gone, killing fish that were never recovered.

The greatest value for halibut is in the fresh-fish, high-end restaurant trade. But because the fishery lasted only a single day, millions of pounds of halibut were frozen and that extra value was lost. By the early 1990s it was clear that the Alaska halibut fishery was a biological success but economically wasteful.

What are individual fishermen's quotas, the IFQs?

In 1995 the Alaskan halibut fishery switched from open access to what is called an *Individual Fisherman's Quota* (IFQ) system. In the open access fishery the length of the fishing season was adjusted so that the actual catch was as close as possible to the *total allowable catch* (TAC). In an IFQ system the total allowable catch is split among individual boats or license holders. Each fisherman knows exactly what his share is and that he is allowed to catch no more than his IFQ.

IFQs are usually assigned based on the share of the total catch the boat has had in the past, although sometimes other factors are considered, such as recent investment in bigger boats and allowances may be made for poor catches due to illness.

With the catch fixed by the IFQ, fishermen are free to choose when to fish and with what size boat. They can also generally sell or lease their shares to other boats. Then we deal with *Individual Transferrable Quotas* (ITQ). In the Alaskan halibut fishery, shares can be transferred, but to forestall anyone amassing too many ITQs, no individual boat is allowed to own more shares than 0.5% of the total catch.

What are the benefits of IFQs?

The IFQs have ended the *race for fish* among boats. There are no more incentives for fishermen to buy ever bigger boats in order to beat the others to the catch. In an open access fishery the best way to make money is to catch the most fish before everyone else. In an IFQ fishery, making money depends on spending less to catch the fish and bringing in higher quality fish to get a better price.

Most important, the fishermen are safer under IFQs. No longer forced out to sea by a limited season, the boats can stay in port or hide out in a safe refuge when storms come up. Ghost-fishing has largely been eliminated too since there is no reason to abandon costly gear.

As a consequence, IFQ fisheries typically become very profitable, and where the IFQ can be turned into an ITQ, fishermen can sell up and retire.

Most fisheries suffer from too many boats chasing too few fish, making it difficult for fishermen to retire or move to another industry. Their investment in boats and gear is worth very little when there are already more boats than needed. In an ITQ program, however, many fishermen who have a valuable and marketable asset will sell to those who want to stay in the fishery and catch more fish. Through those sales, ITQs are reducing the size of the fishing fleet.

Moreover, the quality of the fish delivered to the docks has improved greatly, and most halibut now comes to market as fresh fish year round, a most delicious and much appreciated consequence of the IFQs. The high value of the ITQ has brought wealth to fishermen whose share is significant. The market price for an ITQ is often much higher than the annual value of the catch.

For example, in 2009 the average landed value of halibut was $2.33 per pound, while the average price for halibut ITQs was $19–$22 per pound. When the ITQ price is 8.8 times the landed price per pound, a fisherman who catches $100,000 worth of fish each year can sell the ITQ for roughly $880,000. It is not uncommon for the value of someone's ITQ to be worth hundreds of thousands or even millions of dollars. ITQs have generated wealth that simply did not exist in the open access fishery because of the economic inefficiency of the race for fish.

What are the negative impacts of IFQs?

Despite the many benefits of IFQ and ITQ programs, not everything is as rosy as it seems. When the fishing fleet shrinks, so does the number of people employed as boat owners, captains, and crew. On the one hand, a smaller fleet is better than too many boats chasing too few fish, but for a fishing community fewer available jobs is a heavy social burden—particularly so when ancillary jobs are also lost in the service industries of shipyards, fuel docks, net yards, and chandlers. The consequences of reduced costs for vessel owners are inevitably reduced sales and employment in industries supporting the fleet.

Some ITQs have been drifting away from the traditional fishing communities to the "outside." Sometimes ITQ owners move into town, more often ITQs are sold to people in town or in cities where access to banks and capital is easier. When the cost of entering a fishery can be in the millions of dollars, people in small and isolated communities have less chance to enter. Perhaps the most controversial aspect of ITQ programs is the very wealth they generate. Fishermen who were granted the initial ITQs reaped all the initial benefits. This can be seen as giving away a public resource to a few who will eventually sell and retire as millionaires. Once all the initial ITQ owners have left the fishery, the new entrants won't be blessed with equal windfalls, but they will enjoy more stable economic returns from less variable catches throughout a fishing season of their own choice rather than throwing the dice in a one-day opening.

The design and structure of IFQ and ITQ programs depend wholly on the final objective.

In New Zealand, for instance, economic efficiency is the desired outcome, employment is left to the market. ITQs are increasingly owned by a few large companies or non-fishing investors. Those who are actually on the water catching fish are typically not invested in the ITQ but work for salaries or get paid by the pound caught. Investing in ITQs is just as much a risk as investing in any other asset.

Alternatively, in Alaska there is a deliberate policy to keep the ITQs with on-the-water fishermen. A new buyer of an ITQ must be on the boat that catches the fish. This prevents the New Zealand type of shore-based investor.

What is economic overfishing?

Economic overfishing is more fishing pressure than would make the most profit.

Fishing pressure is the number of boats multiplied by *fishing effort*, that is, the number of days boats fish or the total number of sets of the net or the number of hooks put out. There is a level of fishing effort that will produce the best long-term catch from a population or an ecosystem; any more effort is *yield overfishing*. Similarly there is a level of effort that will bring in the biggest profit; any more effort is *economic overfishing*.

In general, fishing for profit requires less effort than fishing for highest biological yield. Profit from a fishery is the difference between income and costs. Income depends on catch, costs depend on effort. Therefore, if we reduce fishing effort to 10% below the level that will produce highest biological yield, costs will also go down 10%, but revenue will decrease by only 1%–2%. Fishing a little less will be more profitable in the end. Exactly how much less differs from stock to stock and from gear to gear.

Traditionally, halibut prices go up when the catch goes down.

There are added benefits to fishing for profit. Less fishing means less environmental impact, larger and possibly more stable stocks, less by-catch, and less impact of fishing gear on sensitive habitats.

How economically efficient are world fisheries in general?

The short answer is, not very efficient.

In 2009 the World Bank and FAO, the Food and Agriculture Organization of the United Nations, issued a joint report titled "The Sunken Billions: Economic Justification for Fisheries Reform." They estimated that in 2004, 75% of world fisheries were fished too hard at a cost of $50 billion per year in lost profits. These lost billions are equivalent to 64% of the landed value of all fish in that year. Essentially, most of the potential economic value of fisheries is being wasted with too much fishing effort and too many subsidies. Subsidies alone amounted to $10 billion in 2000. Most of that went to cheaper fuel.

How do we prevent economic overfishing?

Eliminating the race for fish as well as the subsidies that encourage more fishing effort goes a long way to stop economic overfishing.

As soon as the number of people allowed to fish is limited and catch shares are allocated among them, the race for fish is over. Allocation can be made through an IFQ system or through giving fishing privileges to entire communities or groups who themselves have developed socially acceptable ways to allocate fish internally. Subsidies tend to increase fleet size beyond best economic outcomes. Without subsidies fishing fleets tend not to grow so much.

Nothing could have been more important than the creation of the 200-mile exclusive economic zones (EEZs) that gave nations exclusive control of fish resources within 200 miles of their coastline. This was a true hallelujah moment; finally, there was a legal framework to stop the race for fish. So long as foreign fleets could pillage anyone's coastal waters with impunity there was no hope to reduce fishing pressure. If there was any profit to be had, someone would come and fish, and this remains precisely the problem beyond the 200 miles to this day.

In 1968 Garret Hardin argued in his paper, "The Tragedy of the Commons," one of the most influential scientific papers ever written, that as long as resources like fisheries were common property they were destined to be overexploited. The advocates of IFQs and ITQs believe that allocating fishing rights or privileges to individuals solves the tragedy of the commons. Yet critics argue that this is privatizing public resources for the benefit of a few.

Are there ways to prevent the tragedy of the commons without privatizing fisheries?

Elinor Ostrom received the Nobel Prize in Economics in 2009 for her work on just this subject. She showed that many common property resources were well managed by communities working together to identify appropriate levels of exploitation. She found that when communities had certain characteristics, such as particularly strong leadership, social cohesion, and exclusive access to resources, they could avoid the tragedy of the commons and find sustainable ways to manage them. As part of his Ph.D. program, Uruguayan Nicolas Gutierrez studied nearly 200 "co-managed" fisheries where communities of fishermen had a significant role in the management. His results support Ostrom's work.

What are community development quotas?

Community development quotas, or CDQs, are used in Alaska to allocate some of the wealth from fisheries to local communities.

When the fishing privileges were being allocated to fishermen based on historical catch, only 92% of the TAC was allocated to historical fishermen and 8% was reserved through CDQs for coastal communities who had a choice of catching the fish themselves or leasing the right to catch them. In the offshore pollock, crab, and Pacific cod fisheries the communities have chosen to lease the quota to fishing companies that own most of the 92%. This provides income for the communities that sometimes stipulate that the leasing company must hire local villagers to work on the ships. In the halibut fishery the CDQ groups often use the quota for locally based boats.

In a twist on the CDQ provisions, the revenue from leasing cannot be directly distributed to the communities but must be used for "fisheries-related" activities. To comply, the villagers have used much revenue from CDQs to buy ownership shares in the same companies that are leasing the quotas from them. As a result, quite a few community groups in Alaska are significant, and even majority, shareholders in the large fishing companies.

How does sector allocation work?

Sector allocation is a more generic form of IFQs in which a share of the catch is allocated to a specific group rather than to individuals. This group must then work out an internal allocation system. They can choose to have an internal race for fish or create an internal IFQ type system.

A large sector allocation system has been in place for the at-sea-processing fleet of factory trawlers fishing for Eastern Bering Sea pollock. They were allocated 40% of the total catch and worked out an internal allocation system that functions very much like an IVQ, an *Individual Vessel Quota*, in which each boat or company is given a specific share of the catch that is allocated to that sector of the ocean in which they are licensed to fish. This has been spectacularly profitable. During the earlier race for fish the season was short, on-board factories were jammed, and then boats sat idle for many months. Now far fewer boats work a much longer season, and because they are not racing each other, the amount of usable product for each ton of landed catch has almost doubled. Profits have soared accordingly.

In 2010 large-scale sector allocation was implemented in New England. Groups of fishermen from the same port using the same gear could apply for a share of the TAC based on their group's historical percentage of the catch. It is far too early to tell how well this system will work there.

What other mechanisms have been used to allocate fish?

Territorial user rights to fish (TURFs) are a long-standing traditional form of management whereby local communities have exclusive rights to an area. Chapter 11 on small-scale fisheries describes how this has been implemented in Chile.

Many economists argue that the state should auction the right to catch fish in the same way oil and gas leases and leases for electromagnetic frequencies for television and cell phones are auctioned. Instead of giving away fishing rights in IFQs, they would be auctioned. The state of Washington auctions the catch of geoduck, a valuable large clam. The state receives about $10 million in revenue each year and spends about $2 million on research and management. It is perhaps the only commercial fishery in the United States where revenues from access exceed the cost of managing it. Working fishermen universally oppose any such system. They are usually in financial trouble when the proposals are made and cannot see how having to pay for the opportunity to fish would improve their situation.

Alternatively, a combination of IFQ programs and auctions has been put forward in which fishermen are given an IFQ initially, but after some time, perhaps 10 years, some portion of the IFQ reverts to the state to be auctioned. This would provide a major financial incentive for fishermen to switch to the IFQ program but over the long term would return much of the wealth to the state. To my knowledge nobody has yet taken that particular bait.

6

CLIMATE AND FISHERIES

How does climate affect fish populations?

Herring are among the most abundant fishes in the world and have been the basis for regional and national economies for centuries. In 1855, over 95,000 people were employed in the herring industry in Scotland. In an 1864 monograph on herring, the British scientist John Mitchell quoted Cuvier: "The coffee bean, the tea leaf, the species of the Torrid Zone, and the silkworm, have less influence on the wealth of nations than the herring of the northern seas. Luxury and caprice may seek those productions, but necessity requires the other.... The greatest statesmen, the most intelligent political economists, have looked on the herring fishery as the most important of maritime expeditions. It has been named the Great Fishery." Between 1950 and 2000, the catch of Atlantic Herring was over 100 million tons, or roughly 5% of all the fish caught worldwide.

There are two very large herring populations in Europe: the Norwegian spring spawning herring, now numbering about 12 million tons and the North Sea herring, now at about 1.5 million tons. The North Sea herring now predominantly spawn from July to October off the northeast coast of Scotland, but historically they spawned on a wide range of banks throughout the length of the North Sea. They attach their eggs to the seafloor, primarily on coarse gravel and small stones. When the eggs hatch ocean currents transport them to the eastern North Sea and the Skagerrak/Kattegat, where they grow for up to two years.

Herring and their relatives are well known for major fluctuations in abundance. The best way to measure the health of the stock is by the total weight of spawning individuals, called *spawning stock biomass*. The biomass of the Norwegian spring spawning stock was 14 million tons in 1950; it declined to just a few hundred thousand tons in the early 1970s and had rebuilt to 12 million tons by 2008. The North Sea stock was over 2 million tons in the 1960s, declined to under 100,000 tons in the late 1970s, and is now back to about 1.5 million tons. These nearly 100-fold fluctuations in abundance are seen across many herring, sardine, and anchovy populations.

The survival of the larval and young juvenile herring depends on ocean currents and food. Herring time their spawning to the annual "bloom" of the microscopic plants and animals that are essential to the survival of the larvae. When the timing is right the larvae find plenty of food and many survive. But when the timing is wrong the larvae find little food, they grow too slowly, and few survive.

We all know that weather is inconstant. A summer may be much too warm and then again way too cold. A winter may be too mild or gruesomely brutal. But we can always hope that next year will be better than the last. Lately, we have become familiar with a weather phenomenon called *el Niño*, associated with warm years in the eastern Equatorial Pacific and, as we are discovering, affecting much of the global climate. Over the last 20 years climate scientists, oceanographers, and fisheries scientists have found that different ocean conditions can last for several decades. The first one popularized was the *Pacific decadal oscillation* (PDO). It occurs in the North Pacific and has two phases: a positive phase, during which the waters along the west coast of North America are warmer than average, and a negative phase, when the warm water shifts to the western Pacific and water temperatures fall below average along the U.S. and Canadian coasts. There was a negative phase in the North Pacific from the early 1950s to the late 1970s, and there has been a positive phase ever since then. This is not to say that every year is warm or cold, but that these conditions run in long periods dominated by specific weather patterns that drive oceanic currents and circulation—and, consequently, the conditions that fish face.

The PDO was first identified because of dramatic increases in salmon runs in Alaska in the late 1970s, and it has now been documented from lake cores that these multi-decadal changes in salmon abundance have been going on for hundreds of years. In the North Atlantic there are two related phenomena: the North Atlantic Oscillation (NAO) and the Arctic Oscillation that control the location of two major weather drivers—the Icelandic low pressure system and the Azores high pressure system. These weather patterns, in turn, determine wind and ocean currents. The resulting multi-decadal changes in ocean conditions are thought to cause periods of high and low survival of herring populations, although in general we cannot directly relate specific environmental conditions to survival in any individual year.

One of the longest running debates in fisheries management is whether spawning biomass is more important than environmental conditions to the survival of juvenile fish and to the ultimate number of young entering a fish stock, often called *recruitment*. The "climate school" would argue that recruitment is largely beyond the control of fisheries management, whereas the "recruitment overfishing school" argues that how many young fish survive depends primarily on how many eggs were laid in the first place. When a stock is overfished, few fish live long enough to spawn more than once and some don't get to spawn at all. When a string of bad years comes along, there are not enough spawners left to replenish the population. If you want to understand why stocks are low, look to overfishing.

Support for these two schools has waxed and waned over the last half century. Through the 1980s the climate school largely held sway because there seemed to be very little relationship between spawning stock and recruitment. However, in the 1980s and 1990s, because so many stocks had been fished so hard, it became clear that when spawning stock went down so did recruitment, and by the late 1990s the recruitment overfishing school ascended to prominence. Most modern definitions of overfishing take this into account. If the stock is below some specific "overfishing threshold" the recommended management action is to dramatically reduce fishing pressure. What this really says is that climate and fishing act together. What seems a safe level of fishing in good times can be disastrous when times are bad.

By 2010 a truce emerged between the two camps. There is now overwhelming evidence from herring and many other species that climate does indeed significantly affect recruitment. Many stocks that were at historically low abundance in the 1990s and early 2000s at the same time produced high or even record high recruitment because of favorable climate conditions. On the other hand, when recruitment is low because of the effects of climate, the total stock biomass declines also. And then we have the paradox that what seems to be low recruitment because of low spawning stock may in fact be the other way around. Low spawning stock may be the result of low recruitment. At the same time, common sense and good ecological work on many stocks have established that it is prudent to maintain fish spawning biomass above some critical level, and almost all fisheries management plans attempt to do this now as insurance for future good recruitment.

The status and management of the Norwegian spring spawning herring and North Sea herring reflect this balanced view. The declines are thought to be climate induced but severely aggravated by excessive fishing. Current fishing plans for both stocks have lowered exploitation rates and specified limited target levels of spawning stock biomass.

Are many fisheries affected by climate?

We expect that all fisheries are somewhat affected by climate. Should we then consider climate changes in the recent past and near future when managing a fishery? Traditionally, fisheries management agencies have ignored climate variation, considering year-to-year variability in productivity to be the major factor in abundance. But once the impact of the Pacific decadal oscillation on fish populations was recognized, there was a big push to think of *decadal long shifts* in productivity rather than simple year-to-year variation.

Current research suggests that these regime shifts may be important in more than half of fish stocks. How does this affect management? Considering regime shifts, we can harvest a higher percentage of the stock during good regimes than during bad ones. However, when a bad regime is very unproductive, all harvesting should stop to save the "capital" for better days. If you have few fish when a good regime starts, it will take a long time to rebuild the stock.

Are there other fisheries where we can look at hundreds of years of history before fishing started?

In 1969, Andrew Soutar and John Isaacs, two biologists at the Scripps Institute of Oceanography in California, published a paper that revolutionized our view of fisheries populations. They found that at certain places in the coastal seas they could take a core of the ocean floor and find historical deposits of fish scales. These special places, called *anoxic sediments*, are found where there is not enough oxygen for bacteria to decompose the scales. Like an archaeological dig, the deeper they looked in the ocean floor the farther back in history they could see the scales of the fish. This allowed them to calculate how many fish of different species were alive over time.

The most interesting result of their work is that the California sardine, which had collapsed in the 1950s off California, had a 2,000 year record of abundance that showed periodic increases and declines long before industrial fishing had arrived. The 1950s collapse was just one of many collapses they could document, although it may have been more severe than earlier ones. The method used by Soutar and Isaacs has been repeated in a number of other marine systems for very abundant sardines and anchovies, and the results have always shown boom and busts similar to those seen in the records of landings in Europe and elsewhere.

A similar technique, this time using nitrogen isotopes, has been used to look at the history of abundance of one species of Pacific salmon in dozens of lakes around the North Pacific. In those studies the Pacific decadal oscillation can be clearly seen going back hundreds of years. While here the highs and lows are not so pronounced as with sardines and anchovies, this technique again gives us insight into abundance and natural variations. The overall message is that climate matters.

How can we tell if a fishery is declining because of climate or fishing pressure?

Fisheries managers around the world scratch their heads over this question whenever a population is in decline. Unfortunately, we rarely understand exactly how climate affects fish productivity, and simple statements like "temperature has risen by 2 degrees, therefore survival will be down by 5%" are impossible to make with accuracy. The problem becomes especially thorny once there is a shift from a productive climate regime to an unproductive one since the population normally decreases with or without fishing. How can we explain periods of high productivity and high population size followed by low productivity and low population size? Did fishing pressure reduce population size, leading to lower productivity—or was it the climate that reduced productivity and then population size? Or, to confound us even further, was it a combination of the two unhappily coinciding? Until we really understand how climate affects fish productivity, we are condemned to wait for the answer for many years. There is general acceptance now that fish productivity is a result of both climate and overfishing. Prudence now dictates that regardless of the relative importance of each of these, when stocks go into decline we need to conserve enough of the spawning stock to take advantage of good conditions if they come back.

What are going to be the impacts on fisheries from a warming ocean?

There is no question that the ocean has, on average, been getting warmer, and this is affecting fish. In the northern hemisphere we find many species farther north every year, and we can already document that some fish do better with a warming climate whereas others suffer. It would be naïve to think that the fish could simply move closer to the poles and all will be well. Obviously fish can move north and south much more easily than plants and land animals, but their well-being is utterly dependent on their food (that is, the primary productivity of the ocean) and quite often their accustomed habitat. The productivity of many species depends on both habitat and food, and if the food moves toward the poles but there is no appropriate physical habitat for the fish, they will not be able to follow. *Ocean primary productivity* is a complex interaction between ocean currents and the ocean floor, especially the location of continental shelves and where there are *upwelling systems* that transport nutrients that become the basis for primary productivity from the deep ocean to the surface.

Niels Bohr, the Nobel Prize–winning physicist, said that "prediction is always difficult, especially about the future." While predictions are being made about the impact of climate change on fish populations, they are suspect and we must reserve judgment. The key to fisheries management will be to anticipate and adapt to changing ocean productivity instead.

What will be the impacts of ocean acidification?

From the perspective of overfishing and fisheries management, ocean acidification is the most frightening aspect of climate change. The ever increasing carbon dioxide in the atmosphere gradually leads to a more acid ocean. We have already learned that even very minor changes in acidity affect the ability of large and small organisms to form shells. There is good theoretical and laboratory evidence that everything from highly desirable oysters, crabs, and corals to the tiny coccolithophores and foraminifera that form much of the base of the ocean food chain may not be able to make shells in a more acidic ocean and thus will not survive.

It is likely, on the other hand, that some species will fill the role of those condemned to oblivion and use the sunlight to take their place in primary productivity, and this productivity will likely find its way up the food chain. This is called "good news" in a dismal scenario. Yet we have no idea what sort of species will fill the niche and what its impact on ocean food chains will be. The oceans and the fish species in a more acidic regime may be totally different, perhaps not even good enough to grace our dinner plates.

7

MIXED FISHERIES

Do fisheries catch one species or more?

Quite a few fisheries catch only one species. Many of the major fisheries fall into that category, including Peruvian anchoveta, the largest fishery in the world, as well as Alaska pollock, the biggest American fishery, and the big European herring fisheries. But many other fisheries catch a mix of species, some that are abundant and productive and some that are under threat and need protection. This makes it much harder to determine the right amount of fishing pressure.

The North Sea trawl fisheries are a typical kind of mixed fishery. Boats from Holland, Denmark, England, Germany, Norway, Scotland, Wales, France, and Belgium fish there for the most important species—cod, haddock, plaice, and saithe. They use trawl nets that are dragged through the water. Fish enter the mouth of the net and are kept from leaving the end of the net, called "cod end," by the flow of the water.

Such fisheries are of particular concern and interest to environmental groups because they catch a mix of species and because of their impacts on the seabed. The North Sea is one of many areas of the world that is intensively trawled. The mid-1990s saw more than 2 million hours of trawling there a year. However, the trawling pressure is not even; some areas are trawled so heavily that a net passes over each point several times a year, while others are hardly touched.

Limited trawling from sailing ships started around the 14th century. But once steam engines came into general use in the 1890s, large-scale trawling took off, and as early as 1900 the fish stocks in the North Sea were a concern. As Walter Garstang from Oxford University observed in 1900, "We have, accordingly, so far as I can see, to face the established fact that the bottom fisheries are not only exhaustible, but in rapid and continuous process of exhaustion; that the rate at which sea fishes multiply and grow, even in favourable seasons, is exceeded by the rate of capture."

The sea floor of the North Sea is now a very different place from what it was 100 years ago, and the North Sea of 1900 was equally different from the North Sea of 1800. René Taudal Poulsen, an environmental historian who examined the changes in ling and cod from 1840 to 1914, showed that ling had been quite abundant in the 19th century and have been rare ever since. Combining Poulsen's work with recent surveys it appears that cod were more abundant in the 20th century than in the 19th, and were most abundant in the 1960s and 1970s.

We are fortunate that concern in the early 1900s led to scientific surveys that allow us to compare how the diversity and abundance of fish has changed after 100 years of trawling in the 20th century. The species targeted by fishing turned out to be less abundant now than they were in 1900 and those not targeted are more abundant. The decline of targeted species reduced competition and allowed some other species to increase. It may be that the overall diversity of species has actually increased over the last 100 years but that the number of fish overall has remained relatively constant. Most significant, however, is that on average the targeted fish are smaller now. There are actually very few large fish of any species these days, which means that the average weight of each fish is less, and consequently there is less total biomass of any target species than there was 100 years ago. This is a natural consequence of even well-managed fisheries, but many of the commercially important fish species in the North Sea are well below the level that would produce maximum sustainable yield.

The central problem of mixed stock fisheries is that it is difficult to protect a weak species while catching a healthy one. The net does not discriminate between species that can tolerate more fishing pressure than others or those who may have had a rough time environmentally, no matter how much we would like to protect them.

Cod and haddock are a current example. Haddock are now considered to be healthy and above target levels, while cod in 2005 were well below target abundance and roughly half of the two- to four-year-olds were caught each year. It was felt that a harvest rate of 20%–30% of biomass would allow the cod stock to rebuild. At the same time the haddock stock was in great shape and could be harvested much harder than at only 20%.

But how to catch haddock and not cod? How to catch the healthy stocks and avoid the weak ones?

Three different approaches are being tried in various places. The first is closing "hot spots." Different species often tend to congregate in separate areas. Such hot spots of the weaker stocks can be closed to fishing. This has been tried in the North Sea but it is difficult to say whether it has had any effect since the closures are kept short and the hot spots are kept small.

A second experimental approach is modifying the fishing gear so that haddock are more likely to be caught than cod. Cod tend to dive when a net approaches while haddock swim up. If the bottom of the net's rope is a little above the sea floor, most cod can be avoided. Different modifications in trawl nets have been very successful at avoiding the catch of turtles and marine mammals in other fisheries.

The third approach relies on fishermen's knowledge about where the haddock and cod are and assumes that they will avoid catching cod. But this works only with strong enforceable incentives. A successful western Canadian version limits the catch of each species for each boat. If a boat catches more than its limit of a species it must stop fishing entirely unless it can lease some additional quota from another vessel. For this system to work, onboard observers must record the catch on each boat. Observer coverage of 100% is now required in both western Canada and the western United States. This approach has not been tried in Europe.

But trawls are not the only kind of gear that catches mixed stocks. Any net, hook, or trap is likely to catch more than one species. Purse seines are large nets that encircle schools of fish; once the fish are caught, the lower part of the net is tightened like a purse. In some fisheries they catch only single species that form large schools, like herring. However, when they are set to catch big schools of skipjack and yellowfin tuna, the more valuable big-eye tuna and various other unintended species also tend to end up in the nets. Most hook and line fisheries, whether commercial or recreational, catch a mix of species. At present, this is something that cannot be avoided, and solutions must be found fishery by fishery.

What determines how hard a fish species can be harvested?

In the absence of fishing, the amount that a population's abundance changes from year to year is determined by the number of births, the body growth of individual fish, and how many fish die naturally. The increase in a population from the year before is called *surplus production.* Generally, when fish populations are at low abundance there is no competition for food, the population increases, and the surplus production is positive. To keep a population stable, you can't harvest more fish than are being produced—the surplus production rate. Many European cod populations are at very low abundance due to overfishing, but their rate of surplus production is so high, on average 40% or 50%, that they can still be fished quite hard without the stock collapsing completely. If harvested at a rate much less than the 40% or 50% a year, those populations will start to increase, and some have indeed done so in recent years. By contrast, many sharks and rays have very low birth rates, and their surplus production rates as well as their sustainable harvest rates may be only a few percent a year.

How do we balance harvesting high- and low-productivity species in mixed fisheries?

If a fishery catches both a highly productive species and one with low productivity, there are two strategies to determine how hard to fish. Harvesting hard enough to get the best possible yield from the high-productivity species will cause the low-productivity one to decline and possibly go locally extinct (it is likely that this happened with many unproductive species in the very early days of fishing). If we want to keep the most unproductive stocks at high abundance we have to fish the productive stocks very lightly, but in doing so we forgo a lot of potential yield from the ecosystem.

We have estimates for many mixed fisheries that fishing to maximize the total yield will cause on average up to 30% of the species to be depleted to very low abundance. However, when that fishing pressure is reduced by half, the loss in yield is actually very low, perhaps only 10%–20%. Consequently, many fewer unproductive species will be depleted, average abundance of all species will go up, and so will profits.

There is no single answer to how hard we ought to fish the mixed stock fisheries, but for many reasons, fishing less hard than at the level of maximum sustainable yield across all the species seems highly desirable.

What is "underfishing"?

Underfishing means fishing less than you could to sustainably produce the most yield or profit.

It is not so different from overfishing because both produce less than maximum yield: one because you fish too hard and the other because you don't fish hard enough.

Economic underfishing can be a choice made to achieve the most possible profit. As we saw earlier, fishing for maximum profit generally means underfishing the yield. It is also possible to underfish the yield but still overfish for best profit. There are sound economic reasons to underfish yield, and since all fishing has ecological consequences, a society may well choose to economically underfish as well.

Is it better to give up potential yield of productive species to keep unproductive species at high abundance?

There is neither a straightforward nor a science-based answer. As we saw earlier, fishing less hard than at the level of maximum sustained yield can be more profitable and have less environmental impact. Certainly in the United States, lawmakers have decided that keeping all stocks viable is desirable, and there are legal requirements to rebuild any depleted unproductive stocks. Other societies may make different choices depending on their emphasis on food production, profitability, ecological impact, or jobs.

How can we manage fisheries to reduce the mixed nature of the fishery?

The simplest approach is to fish less hard and reduce impacts on the less productive species. Earlier we discussed the potential for area closures, gear modification, and use of incentives to individual boats to find ways to catch productive species and avoid unproductive ones. Yet another approach is to completely change fishing gears. Fish traps, for instance, tend to be more selective than trawl nets, and even various forms of hook and line may be more selective. But changing fishing gears can be tricky for management agencies because the catch is often allocated to specific fishing gears. Allocating catch to different types of gear means taking the fish away from some people and giving it to others. Imagine being a trawler and suddenly losing your share of the fish and thus your investment in gear and income. Of course you would fight through all available legal and political channels.

My own opinion is that it is best to provide direct incentives to the fishing fleet by assigning them collective or individual shares of the catch, let them find ways to catch their share of each species, and give the fleet the choice of where to fish, when to fish, and what gear to use.

8

HIGH SEAS FISHERIES

What is the status of bluefin tuna that were proposed for CITES listing?

Throughout the western Mediterranean Sea, an annual ritual known in Sicily as *matanza* has taken place for perhaps 400 years. Nets are set in shallow water near the shore to form a funnel, a giant trap to catch bluefin tuna as they migrate past, herding them into a series of ever smaller open chambers to the final "chamber of death." For many coastal villages, the *matanza* meant far more than a hunt for meat. It was the trap, the cooperative that went out to hunt, kill, and butcher the tuna, and it was the soul of the community itself. But the *matanza* in Sicily is no more, the nets lie rotting on the shorelines, the young men have gone to look for employment in the cities, and the culture that tied the people to the sea and to the bluefin tuna is almost completely dead.

Bluefin tuna, torpedo shaped, very fast, and weighing up to over 1,000 lbs, are some of the most impressive animals of the sea. They are also the modern poster children of overfishing; many argue that they are on the verge of extinction. In 2010 they were proposed to the *Convention on International Trade in Endangered Species of Wild Flora and Fauna* (*CITES*, pronounced sigh-tease) for listing in Appendix I, an act that would have made it illegal to sell Atlantic bluefin tuna products internationally. They are some of the largest fish of the ocean, and to a biologist they are one of nature's greatest creations, warm-blooded in a cold ocean, able to swim at great speeds and travel thousands of miles in fantastic migrations. To lovers of sushi, they are among the most prized fish in the world; their fatty flesh, particularly the belly flap, is the ultimate delicacy, especially in Japan, where most of the bluefin tuna of the world is eaten.

Atlantic bluefin spawn in the Mediterranean Sea and in the Gulf of Mexico. The two populations are genetically distinct, but they do mix during their migrations. The juveniles of the Mediterranean stock largely stay put for several years before they begin to migrate. They mature as early as four years of age and live to be at least 20. There is evidence that the Gulf of Mexico stock spawns at an older age. Because of their size and speed only the largest and fastest of predators, such as killer whales and large sharks, are thought to be a threat.

Up to 1900, the *matanza* of the Mediterranean was the only significant fishery, until a sport fishery developed around 1900. In the western Atlantic there was no market for bluefin; their meat was considered too fatty until the export market to Japan opened up late in the 20th century. After World War II, Japan began a longline fishery on the high seas for large tuna and billfish, with bluefin tuna the most prized. This fishery grew rapidly in the 1950s. High seas catches of the Mediterranean stock peaked at about 30,000 tons in the 1950s and have since declined to between 5,000 and 10,000 tons a year. The big change, and the big threat, has been the expansion of fisheries in the Mediterranean itself, especially those using large purse seines that can capture entire schools. The catches rose from 5,000–10,000 tons annually by 1970, and by now up to 30,000 tons are taken each year. These catches are particularly significant. Since each juvenile weighs much less than an adult, the numbers of tuna caught have grown greatly. Most of the adult fish are now shipped to the fish market in Tokyo; the juveniles are often captured live, towed to large pens near the coast, fed to increase their weight, then killed and also sent to Japan.

Of the two stocks of Atlantic bluefin, the much larger eastern stock is of the most concern. While the level of depletion of eastern Atlantic bluefin tuna is uncertain, it is very clear that the catch is too high and abundance too low. The *International Commission for the Conservation of Atlantic Tunas (ICCAT)* reported to CITES in 2009 that the stock was between 20% and 70% of an abundance that would produce maximum yield, and that the current harvests, allowing also for illegal and unreported catches, were two to five times too high. Independent scientific assessments show similar results. There is no doubt that the stock is overfished and the relative harvest rates are much too high. Immediate reduction in catches is urgently needed.

Even though the current abundance is not really "near extinction," extinction is a likely outcome if the current harvest rates are continued. Many other stocks are at lower relative abundance, but Atlantic bluefin stand out because of the high harvest rates. A particular threat to them is the purse seine fishery in the Mediterranean. If there continue to be fewer and fewer bluefin to catch, high-seas long-lining will no longer be profitable and fishing pressure will probably drop long before the stock will be anywhere near extinct. However, because purse seines catch the fish when they are packed together in schools they may well be able to catch the very last one.

The International Commission for the Conservation of Atlantic Tunas is one of five international fisheries organizations responsible for tuna management on the high seas. Each of these organizations, and ICCAT in particular, is relatively ineffective for the same reasons. First, they normally operate with the requirement of either a complete consensus of members or a super-majority to adopt any management regulations. This means that whatever conservation measures they might be willing to adopt, they are limited by the most reluctant member nations. Second, each country is usually responsible for data collection and enforcement on its own vessels. As a result, compliance—with already often lax management measures—is poor. Major over-catches continue to be found. Unsurprisingly, conservation NGOs refer to ICCAT as the International Commission to Catch All the Tunas. On the whole it is hard to make a case that the international tuna organizations have had a significant impact on the actual catches. It is not clear that we would be in a much different state than we are now if there were no high-seas management at all.

What is the status of tuna around the world?

In 2003, *Nature* published an article entitled "Rapid Worldwide Depletion of Predatory Fish Communities" that made the front pages of newspapers around the world; it proclaimed that all the large tunas of the world had been depleted by 80%, as early as 1980. Although this conclusion has been rejected by every scientific body that has examined the data, the perception that the tunas of the world are highly depleted remains deeply rooted in the scientific community.

The reality is that as of 2010, the large tunas of the world are, overall, at about 50% of their abundance compared to their status when large-scale industrial tuna fishing began. This figure is above the management targets. Only the bluefin tunas are overfished. There are five major types of tuna caught on the high seas and regulated by international agencies. In order of size and value of individual fish they are bluefin, big-eye, yellowfin, albacore, and skipjack. Bluefin and bigeye, along with higher quality yellowfin and albacore, are sold almost exclusively for sashimi and sushi, while the lower quality fish, including almost all skipjack, are canned.

Bluefin are universally overexploited. As we have seen, fishing mortality remains much too high in the Atlantic, but even more depleted than the Atlantic bluefin is the southern bluefin tuna stock in the Indian Ocean. Catches there have been reduced significantly yet they are still above the level that would produce *maximum sustainable yield* (MSY). None of the other tuna stocks would be classified as overfished by U.S. standards, where overfished is defined as the point when the biomass of a stock drops to a level that it can no longer produce near MSY. Several of the bigeye and yellowfin stocks have dropped below their long-term target biomass. Most yellowfin stocks are currently right in the target range of biomass and exploitation rate, yet we find a worrying trend of increasing catch and decreasing biomass. Skipjack and albacore are generally in good shape—except for North Atlantic albacore, whose harvest rates are too high.

While, with the exception of bluefin, tuna stocks overall are doing well, few thanks are due to effective management. Rather, the economics of fishing seem to dictate catch limits at present. The future of the major tuna stocks seems to depend more on the price of oil and the market price of tuna.

Have some international fisheries management organizations been successful?

Thankfully, yes. An agency that stands out is the *International Pacific Halibut Commission*, founded in 1922 by only two countries, the United States and Canada. Like other international commissions it relies on consensus and national self-enforcement, but because there are only two members it works very effectively.

The *Commission for the Conservation of Antarctic Marine Living Resources* is generally credited with much success in implementing an ecosystem approach to fisheries management and finding ways to reduce illegal fishing; it does this by registering vessels that are legally allowed to fish and tracking the legally caught product through the world markets.

The *Inter-American Tropical Tuna Commission* can be credited with a very effective program to reduce by-catch of dolphins by modifying the way that the nets are operated.

The International Commission for the Conservation of Atlantic Tunas did succeed in rebuilding the North Atlantic swordfish by catch reductions when this fish was below management targets.

Why are some tuna stocks underexploited and others overexploited?

Money talks. The evidence is that the major driver is economics. Bluefin and bigeye are extremely valuable and heavily exploited. Skipjack is their poor relation and generally well above management targets. One could always argue of course that the high-value species have the lowest sustainable harvest rates because they live so long, but I think the economics argument wins out. Also, tuna are still with us by the grace of their history. Compared to cod or herring, tuna are latecomers to the world of large-scale fisheries. The Atlantic tuna were the first to be fished commercially and tend to be the most intensively exploited. The skipjack and yellowfin of the Indian Ocean were last and there has simply not been enough time to fish them down as much as the others.

Is there hope for managing these high-seas fisheries?

In general, there is not much to crow about. International agencies do not have a laudable track record. Remember that they are not much more than a collection of component member nations, so the question should perhaps be rephrased, "can countries work cooperatively to manage the high seas?" On the whole, not yet. Few are willing to relinquish national sovereignty and routinely allow international observers on board. The management successes that some countries have achieved in their own 200-mile zones will be hard to repeat until national governments let the international fisheries management agencies have more control over regulation and enforcement.

Yet there are some encouraging signs of better compliance through international agreements. The threat of a CITES listing for Atlantic bluefin and the pressure from NGOs seems to have motivated many countries to act more responsibly. But as long as self-interest on the high seas predominates, I remain pessimistic about future management beyond national 200-mile zones.

9

DEEPWATER FISHERIES

What happened to the orange roughy stocks?

In 1987, the director of the primary fisheries research laboratory in Australia, Roy Harden Jones, announced that a research survey for a species called orange roughy had found schools off southern Australia with an estimated total biomass of 1 million tons. At that time orange roughy was worth $2,000 per ton, so Harden Jones had essentially said that fish worth $2 billion were out there waiting to be taken. Calculations suggested that the value of the annual sustainable yield would be as much as $200 million per year at normal sustainable harvest rate. The orange roughy gold rush in Australia was on. Licenses skyrocketed in value, bigger boats were built, and a lot of champagne went down the throats of fishermen lucky enough to already hold a permit to fish in those waters.

But like many gold rushes, the Australian orange roughy fishery too turned out to be mostly fool's gold. The 1 million tons of orange roughy were an illusion. What was reported as orange roughy turned out to be sonar echoes from rocks on the bottom—and no one buys rocks for thousands of dollars a ton. The total catch of orange roughy taken over the years has been fewer than 300 thousand tons.

Even more surprisingly, orange roughy did not turn out to be a "normal" fish with normal rates of sustainable harvest. Instead it is one of the longest-lived fishes in the world. Instead of being able to be sustainably harvested at perhaps 20% a year, the sustainable harvest rate is at most at a few percent a year. Even now, almost 25 years later, we still don't really know what percentage of orange roughy can be sustainably harvested. The Australian orange roughy fishery is now essentially closed. A few small fortunes were made during the initial rush but the dreamed of fabulous wealth proved elusive.

Orange roughy are found in both the northern and southern hemispheres in deep water; they are often associated with under-sea mountains, but some of the largest catches come from deep flat bottom. They grow slowly and reach their full size of 35–60 cm after 20–40 years. They do not reproduce until they are about 30 years old and commonly live to be 100. They are most vulnerable to fishing when they get together in massive aggregations to spawn. About the rest of their lives— where they live, how far and when they move—we know very little. Genetic studies and differences in size suggest that many populations may be isolated. It may well be that each spawning aggregation is reproductively different, much like salmon.

A World Wildlife Fund website proclaimed, "Reckless overfishing is rapidly causing the demise of orange roughy," a comment typical of the environmental NGO concerns. "Orange roughy fisheries are a testament to unsustainable fishing," says the Greenpeace website. Every list of what to eat published by environmental groups puts orange roughy into the "do not eat" category, and Greenpeace has been successful at persuading a number of retail chains to stop selling it.

There are many environmental concerns about orange roughy fisheries. All the fished stocks have declined significantly. We cannot measure their abundance reliably and are uncertain how many there are and how far they have been depleted. They are often caught with bottom trawl gear in areas that are very sensitive to trawling. Finally, orange roughy are fished sometimes deeper than 1,000 meters (3,000 feet) and their high value has led fisheries to expand into new, largely unknown and previously unfished areas that many environmental groups would like to see protected from all fishing.

New Zealand was the first country to develop orange roughy fisheries and today remains the dominant supplier. Orange roughy had been known from research cruises for many years. Some foreign trawlers had experimented with fishing very deep by the standards of the day and brought up some high individual catches, but it was not until New Zealand declared a 200-mile zone in 1978 that a significant fleet capable of fishing at the depths of orange roughy developed. The rewards were stupendous. A single 15-minute tow of a trawl net could bring up 50 tons of fish, worth $100,000. Catches rose fast in the early 1980s and reached a peak of over 50,000 tons per year by the middle of the decade—a catch worth $100 million annually.

In the beginning, as the lucrative nature of the fishery became clear, exploratory fishing found more and more orange roughy stocks around the country. In the 1980s, New Zealand instituted a quota system for most of its fisheries, including deepwater species. Catches were controlled from the outset, but overly optimistic assessments of stock size and estimates of the sustainable yield led to ever higher quotas. The quota for the major stock was raised from 23,000 tons in 1983 to 38,000 tons in 1989. When it became obvious that biomass estimates were too high and orange roughy were so long-lived, sustainable yields had to be lowered and the quota for the major fishery is now down to 10,000 tons.

A population with a high rate of natural mortality must, by necessity, have a high rate of replacement. For instance, if 20% of fish in a population die each year from predation, this population must produce roughly 20% of new fish by *recruitment* (births) and growth to compensate or the population will decline. Considerable experience across a range of fish stocks shows that the sustainable rate of exploitation is about the same as the natural mortality rate. Therefore, if orange roughy are indeed very long-lived, it stands to reason that their natural mortality rate is very low. If as many as 20% actually did die each year, there would not be that many centenarians. Consequently, the best estimates are that annually only 2%–6% of orange roughy die, which, in turn, brings down the sustainable harvest rate to no more than a few percent a year.

A combination of this revised harvest rate, the rapidly declining catches, and the results of various surveys prompted the New Zealand government, in the 1990s, to declare dramatic quota reductions for most fishing grounds.

The Challenger Plateau is an undersea plain off the west coast of New Zealand that used to support a significant orange roughy fishery with a quota high of about 10,000 tons in the 1980s. As an experiment, both the government and the fishing industry decided to keep quotas higher than what they thought was sustainable to see the long-term impacts of high fishing pressure on roughy stocks. At the beginning of the fishery, each time a net went into the water it pulled up an average of 15 tons. By 1998 it was down to less than one ton, and when corrections were made for the changes in vessel size and other improvements in fishing technology, it was estimated that only 3% of the original Challenger roughy stock remained.

The experiment had succeeded in demonstrating that it was possible to collapse an orange roughy stock. The fishery was closed in 2000. For many this is the ultimate symbol of the unsustainability of fishing the species.

Orange roughy fishing marked the beginning of a swift expansion of deep ocean fisheries. Most of the world's fish catch comes from either the shallow waters of the continental

shelves or from surface fisheries on the high seas. There was no deep water fishery before orange roughy. But once those high-value stocks were found in the 1970s, the technology to fish commercially at depths of over 1,000 meters developed very quickly.

Modern global positioning systems tell the boat exactly where it is, and electronic equipment on the nets shows depth and location. Today a skipper can place a net almost exactly where he wants it even though there are several kilometers of cable between his boat and the net. In the early orange roughy days, a real problem was that the aggregations were so dense that nets often filled in minutes and could be ripped apart. Today's nets are outfitted with mini echo-sounders that tell a skipper how many fish he has in the net so he can pull it up before it gets too full.

Unfortunately, our understanding of orange roughy biology is not nearly so sophisticated.

We generally measure the abundance of fish in the ocean by either towing nets in a scientifically designed survey or using sonar-like echo-sounders that tell us how many fish are down below. Because orange roughy form such dense schools, traditional net-based surveys are highly variable. Hit a school and you get a lot of fish; miss a school and you get almost nothing.

Echo-sounding is equally confounded because orange roughy stay so close to the ocean floor. Their acoustic signals are hidden by echoes from the bottom, and to add even more frustration, they are fiendishly hard to "see" with sound because if even a few other fish are mixed in they make the school look bigger than it really is. Yet, we need to know the precise fraction of orange roughy in a school to judge their true numbers.

Initial estimates of the sustainable yield of orange roughy were so optimistic because nobody knew how truly old they get. Even though New Zealand scientists used only half the productivity values of shallow species, this was too high for orange roughy. Even after 40 years of research we are still uncertain about their natural mortality rate. It is not only very hard to determine their age, but there may also be large differences in age structure from place to place. We are now confident that the sustainable exploitation rate is low, but we don't really know how low. We also do not know what role they play in their ecosystem nor do we know what impact fishing has over time on the deep ecosystem where they live. But we do know that fishing has an impact on the seafloor because roughy are fished with bottom trawl nets. And how those trawl nets change the ecosystem is at the core of many of our concerns with this fishery.

Can very slow-growing fish like orange roughy be sustainably managed?

Based on everything we know about biology, there ought to be a sustainable yield of orange roughy. It is certainly much less than we originally thought and may well be less than we think now. Even the longest-lived species should be able to be sustainably harvested if we content ourselves with a small enough fraction.

For instance, the geoduck is a very large clam that also lives well over 100 years. New fisheries for geoduck in British Columbia and Washington State have harvested about 1% of the population each year, with a long-term management strategy to take 50% of the initial biomass over the first 50 years of the fishery. Compare this with orange roughy, where up to 20% of the biomass was taken each year and most populations have been reduced to 30% or less of the original biomass.

In retrospect, and with current knowledge of orange roughy biology, it would have been better to take a much smaller portion of the stock each year and develop the fishery much more slowly.

We can also look empirically at how a fish population increases and produces sustainable yield. With one exception, every managed orange roughy stock has declined even after quota levels were substantially reduced, which gives us little empirical evidence that the populations can produce any sustainable yield whatsoever. The one exception is the stock on the Challenger Plateau, which was thought to have been fished down to 3% of its original biomass when the fishery closed in 2000. However, surveys in 2006 and 2009 showed large aggregations of orange roughy once again, estimated to be on the order of 30% of the biomass before any fishing began. The fishery has since been reopened with a very small experimental quota.

What is the experience with orange roughy in other countries?

New Zealand, Australia, and Namibia were the big players. Small catches have come from the northeast Atlantic, Chile, and the Indian Ocean. Australia has closed its major orange roughy fisheries and Namibia's is closed entirely. New Zealand is still trying to determine whether current catch levels are sustainable enough to allow rebuilding.

In almost every case, initial estimates of abundance from echo-soundings and trawl catches suggested large abundances. But as soon as the fisheries got up to speed the fish seemed to disappear altogether, after disappointingly small catches. Either the initial estimates of abundance were far too high or fishing causes orange roughy to swim away.

Does closing large sections of New Zealand's economic zone assure the sustainability of orange roughy?

New Zealand has closed about 30% of its economic zone to bottom trawling and effectively to orange roughy fishing. This includes many sea mounts and much orange roughy habitat, which assures that in New Zealand they are no longer in any danger of extinction. However, because the evidence suggests that the populations are generally discrete and very likely do not affect one another, the closed areas do not contribute to any sustainability of the other historical orange roughy fishing areas.

Should we have left potential orange roughy stocks unfished until we know more about their biology and ecosystem?

We knew nothing about the biology of orange roughy to set any quotas before the fishery started, and that fishery was driven by money chased by both fishermen and governments.

In 1919, W. F. Thompson, one of the premier fisheries scientists of the first half of the 20th century and the original director of the International Pacific Halibut Commission, summarized the fisheries approach that dominated until recent years as "proof that seeks to modify the ways of commerce or of sport must be overwhelming." Few governments had the power or will to "modify the ways of commerce" when millions of dollars were to be made.

Thompson also summarized the biological and management problem: "There is no way of knowing the strain a species will stand save by submitting it to one." In the absence of commercial fisheries for orange roughy and the $70 million of profits spent on research, we would likely never have learned enough about their biology to manage them.

How should we deal with new resources when their biology and sustainability are highly uncertain?

The most reasonable approach would seem to be experimental fishing. We could open up a few fishing grounds to careful development—for example, by sending out only a few vessels to limit capital investment, always combined with intense research, and, at the same time, keep most of the potential fishing grounds untouched or "in the bank." Some of the great wealth that may come from the development of new fisheries could be channeled into the research necessary to understand how the stocks could be sustainably managed. Of course, this would require such painful restraint on the part of regulatory agencies that we have hardly ever seen it.

10

RECREATIONAL FISHERIES

Are recreational fisheries fundamentally different from commercial fisheries?

The simple answer is yes, recreational fishing is very different from most commercial fisheries. Many more people participate, their catch and effort are harder to measure, and the overall objective of recreational fishing is often very different from that of commercial fishing.

First, in recreational fishing, very many catch a few each, whereas in the commercial fisheries, a few catch a whole lot. In the United States, recreational fishing is worth $82 billion in sales and rental of fishing paraphernalia, from boats to life jackets and rods to bait, and it provides 533,813 jobs. The Gulf of Mexico recreational fishery is one of the largest in the United States and indeed the world. From the Florida Keys to South Texas, more than 3 million sport fishermen make roughly 25 million trips each year and account for 40% of all U.S. recreational catch in the ocean.

The Gulf sport fishery can be separated into three categories: shore-based fishing, fishing from private and rental boats, and fishing from commercial charter boats by paying anglers. In the Gulf the catch by private and charter boats is roughly the same while shore-based fishing is much less important. The differences between these groups are great.

Shore-based fishermen often walk to the shore and spend very little on their gear. High-end fishermen for trophy billfish may spend many thousand dollars a day. But anglers in private boats vastly outnumber and collectively outspend all other types of recreational fishermen.

Each kind poses special challenges to management. Measuring the catch, the initial step in fisheries management, is the first great hurdle to any meaningful catch statistics, given the sheer number of anglers. Charter boat catches are by far the easiest to measure and track and are similar to commercial ones for the purposes of research. Their licenses record who they are, they operate out of a few ports, they can carry onboard observers, and they can be sampled with a variety of the usual techniques such as logbooks or sampling by government officials when they dock. Anglers in private boats launch from thousands of individual docks and ramps. Tracking their movements and measuring their effort and success is quite a challenge. Accordingly, there is much uncertainty in the resulting numbers.

The objectives of recreational and commercial fishermen are vastly different. Commercial fisheries are about production of food, income, and employment. Sport fishing is primarily about the experience and calls for different management methods. Management of commercial fisheries is mostly concerned with trying to regulate fishing pressure to achieve high yields, but in sport fishing the objective is almost always to maximize fishing effort. Commercial fishermen and managers are constantly looking for methods to reduce the cost of fishing. In sport fishing, the more people spend, the better. At the extreme, high-end fly fishermen often spend thousands of dollars a day to fish in exclusive resorts, spend again as much on their gear, and release everything they catch.

Another difference is the importance of large fish. Commercial fishermen like large fish because they weigh more and are more valuable. Sport fishermen love large fish, especially trophy size ones, and the sport fishing industry loves them even more. In response, managers often mandate a very high legal size limit to encourage the release of small fish and give them a chance to some day make a trophy angler happy.

Sport fisheries are very frequently mixed fisheries. Several different species may be caught in any place with any gear. If one method of fishing for a certain species is unsuccessful or if the catch limit on a species is reached, fishermen will simply spend the next few hours in a different place angling for a different species. In the Gulf of Mexico, the most prized reef fish is the red snapper, but catch limits are so strict that the season is often just a few weeks long—and snapper are only fifth on the list of pounds landed (in 2006). King mackerel, sheepshead, red drum, and spotted sea trout all weigh in with higher sport landings.

In the Gulf of Mexico, approximately 50% of the fish caught are released. Among the many other challenges faced by managers is trying, for the purpose of population statistics, to figure out how many of those released fish actually survive.

Given that simply going fishing and the thrill of landing a fish are the main objectives of recreational fishing, catch and release is all about effort and landing, with much less impact on target populations. In some sense, catch and release is the ultimate form of sport fishing.

But to this coin too there is a reverse. Several Alaskan native communities prohibit catch and release fishing in their territories because they consider "playing with food" inconsistent with their cultural tradition. Animal rights groups often object to all forms of fishing, but catch and release is particularly offensive to them because they feel that anglers derive pleasure from making animals suffer. Catch and release fishing is banned in both Germany and Switzerland due to concerns about animal welfare.

With the number of people involved in recreational fishing comes considerable political power. While recreational fishing groups in the United States often complain that they are underrepresented on the fishery management councils, there is no denying their political power in state and federal government. Recreational fishermen in Florida succeeded in outlawing many kinds of commercial fishing, and similar measures have been put forward in other states. The income from sport fishing licenses is often a major source of funding for state fish and wildlife agencies and it is, therefore, in their best interest to encourage more sport fishing effort.

What is the scale of recreational fishing in the United States and Europe?

By current estimates, there are about 30 million licensed anglers in the United States, generating $45 billion in annual sales. Surveys indicate that about 60 million Americans call themselves anglers. This represents 20% of the American public and a very large constituency. In contrast, participation in Europe is highly variable, with only 1% of Italians but 40% of Finns involved in recreational angling.

In general, recreational catches are small compared to commercial catches, but for the most desirable species the recreational catch is often a large fraction of total catch. In the United States, recreational catch constitutes only 3% of total landings, but when large industrial pelagic fisheries are excluded it rises to 10%, and when we look only at species that are overfished or of conservation concern (such as red snapper), the recreational catch can often be over 50% of landings.

How does recreational fish management differ from management of commercial fisheries?

The traditional tools of recreational fisheries managers are gear limits, size limits, and time and area closures. While commercial fisheries are increasingly given hard total catch limits, the difficulty of collecting catch data on anglers has meant that this tool is rarely used. Instead, when recreational catch is too high, season lengths and size limits are reduced and daily or seasonal total catch limits are often imposed on each angler.

Measuring recreational catch is difficult. The number of people involved and the many landing places makes sampling programs very expensive to implement. Few management agencies attempt to collect recreational fisheries statistics by the kinds of landing surveys common in commercial fisheries. Charter boats are often monitored by combinations of logbooks, observers, and landing interviews, but for individual fishermen, recreational catch and effort is collected primarily by telephone interviews.

How is management different for freshwater recreational fisheries and saltwater recreational fisheries?

The biggest difference is hatcheries. Almost all management agencies put juvenile fish into fresh water, and increasingly saltwater, to supplement natural production. In many places almost all freshwater caught fish are artificially produced, and many freshwater fisheries are described as exclusively put-and-take. There is little natural production of fish. The political power of recreational fishing is strong enough and hatchery technology advanced enough that the instant response to not having enough fish is to produce more. We see here a natural confluence of objective, effort, and political power to produce more hatcheries.

While hatchery technology is most advanced in freshwater, saltwater hatcheries are not far behind. Already, dozens of marine species are being propagated in hatcheries around the United States to augment the recreational fisheries, although the evidence for success of these activities in saltwater is much more difficult to detect than for freshwater species.

Freshwater anglers were also responsible for the introduction of quite a few exotic species around the world to create new fisheries. Take rainbow trout, for instance, now tempting anglers on every continent but the Antarctic and introduced, not always legally, by sport fishermen.

Does recreational fishing play a role in overfishing?

In marine fisheries, the impact of recreational fishing is very different from fishery to fishery. For large industrial fisheries, recreational fishing is rarely a significant issue. But for the most desired species, recreational fishing may be the greatest source of mortality and is an integral part of the problem of overfishing as well as the solution.

Freshwater recreational fishing is much more important than commercial fishing in the United States and Europe. In much of Asia, Africa, and South America there are still large artisanal freshwater fisheries, but in the United States and Europe recreational fishermen have won the allocation battles and most of the catch is theirs.

Perhaps the biggest impact recreational fishing has had on biodiversity has been the introduction of exotics as well as operating hatcheries in freshwater. By now, many societies have chosen to create new recreational fisheries by rearing exotic species in hatcheries at the expense of their native fishes. Thus, if we consider the declines in native species due to competition with introduced ones under the umbrella of overfishing, recreational fishing has definitely played a major role in much of the world.

11

SMALL-SCALE AND ARTISANAL FISHERIES

Many of the fisheries of the world are small scale—how can they be managed?

The "loco" is a carnivorous marine snail common to the rocky shores of Chile and Peru. A large individual is about the size of a slightly flattened fist and would be primarily of zoological interest except for one thing. The loco are also very tasty. Their meat has been regularly eaten by local people along the coast of Chile since the beginning of human occupation, mostly by collecting them at low tide or diving in shallow water. Prior to 1974 there was a significant small-scale artisanal fishery for local consumption. But in the early 1970s, Chilean economic policy changed, exports were encouraged, boats and processing plants were subsidized, and a market for loco developed in Asia, where the snail was marketed as Chilean abalone. Price and demand grew rapidly as did fishing effort and catch.

By 1980, catch had increased four- to sixfold. The fishery was largely "open access," meaning that fishermen could fish where they wanted, when they wanted. While the loco fishery had traditionally taken place locally, around the hundreds of small fishing communities that dot the Chilean coast, the high prices encouraged many fishermen to search up and down the coast for new untapped loco populations. "Loco wars" erupted when local fishermen tried to protect their traditional fishing areas from outsiders. During the 1980s, catches declined and loco became very hard to find; the government tried a range of traditional fisheries measures, including closed seasons, catch limits, and other methods, but these proved totally ineffective. The fishery was closed completely in 1989.

Along 4,100 km of Chilean coastline there are 425 small fishing communities known as "caletas." In each caleta fishermen are organized in formal associations known as "syndicates." Caletas are almost always associated with a specific landing point for boats—the word "caleta" means a cove or small bay in Spanish. The fishermen of caletas are artisanal fishermen. Because they sell their catch they cannot be considered "subsistence" fishermen, and yet their boats are small enough that they do not qualify as an industrial fishery. Each caleta fishes its local resources, including benthic invertebrates that can be harvested by collecting or shallow diving and fish caught with longlines or purse seines. The diversity of what they harvest is impressive. Loco is generally the most important, but clams, barnacles, seaweed, crabs, limpets, and sea urchins are all common. It would not be unusual for a caleta to harvest 20 different species. The collapse of the loco fishery hit the caletas very hard—it had been the major source of income—and in these small fishing communities there were few other opportunities for employment. Hundreds of caletas were looking for ways to sustainably manage their local resources because their very existence depended on finding sustainability.

Chile, like many countries of the world, has largely adopted a Western style "top-down" management system. There is a centralized fisheries agency that coordinates data collection and research, sets regulations, and has enforcement officers to try to assure compliance with these regulations. This "top-down" system is designed for large-scale industrial fisheries where there is a single stock of fish whose abundance can be determined and where the number of fishermen and landing ports is small enough that the catches can be monitored. But these assumptions do not hold for the artisanal fisheries of Chile. The species being fished are largely sedentary, which means that the size of the population in one caleta may be very different from the size of the population a few hundred kilometers down the coast at another caleta. Invertebrates are often regulated by prohibiting the landing of individuals below a minimum size. This size limit is typically set big enough that some individuals are allowed to reproduce before capture. But with sedentary species the growth rates can differ greatly from place to place, and the appropriate size limit in one part of the coast, or even one side of a rocky reef, may be different in another place. Regulations and management need to be very locally adapted. Finally, Chile simply did not have the resources to enforce its regulations across the hundreds of individual caletas spread along its 4,100 km of coastline. While the loco fishery was officially closed in 1989, a considerable illegal trade continued, and local fishermen were powerless to prevent continued fishing of their local stocks they hoped would rebuild.

Juan Carlos Castilla is a marine ecologist and a professor at the Catholic University of Chile in Santiago. His university has a marine laboratory on the coast west of Santiago, and in 1982, Castilla convinced the local caleta to let him set aside a portion of the rocky shoreline at the marine laboratory as an area closed to fishing. The results for loco were dramatic. Within two years loco were abundant and large in the area protected from fishing, whereas just a few hundred meters down the coast where fishing continued, loco were rare. This small-scale experiment provided clear evidence that it was fishing, not bad environmental conditions, that had led to the collapse of the loco, and that loco could be managed on a very small scale.

In 1991, Chile introduced a new fisheries law that allowed for the formation of MEABRs (Management and Exploitation

Areas for Benthic Resources) that permit fishermen's organizations from the caletas to apply for the right to exclusive access to tracts of seabed or regions of the coastline and to manage the *benthic resources* (plants and animals of the ocean floor) of that area under a co-management regime. The caleta is responsible for formulating a resource inventory and developing a management plan. The central government evaluates the plan and monitors its implementation. Most important, the caletas have the legal ability to exclude anyone else from exploiting benthic resources within their assigned areas.

The system has generally worked well. In 2005, there were 547 registered MEABRs encompassing 102,338 hectares. Within the areas managed by caletas, loco are much more abundant, and given the security of exclusive access, each caleta can develop a business plan to maximize their income from the various resources under their control. Abundance of resources has increased, incomes have increased, and the members of the caletas feel empowered and more in control of their own destiny. There are certainly problems; some caletas find that their MEABRs are too small and the natural variability of good and bad years within their small area is too high. The quality of governance and perceived fairness of allocation of fishing opportunities and income within caletas is highly variable. But by and large, this system of territorial fishing rights is viewed as a success and a model for how small-scale, sedentary resources can be managed worldwide.

Is Chile typical of small-scale fisheries?

The situation of the caletas in Chile is quite typical. Most small-scale fisheries rely on a broad range of resources, many of which are sedentary. The "top-down" mode of management is totally inappropriate for these fisheries because central governments will never have the resources needed to understand local conditions nor the ability to enforce regulations in hundreds of small-scale communities. What is atypical of Chile is the legal framework it has put in place and the willingness of the central government to divest power to local communities. A somewhat unique feature is the low population density along most of the Chilean coast and the very discrete nature of the caletas. Moreover, the caletas provided a pre-existing institutional framework for establishment of territorial fishing rights.

How were fisheries managed prior to modern governmental fisheries agencies?

Bob Johannes was a marine biologist who spent many years studying the traditional management of fishing communities in the western Pacific, and his 1981 book *The Words of the Lagoon* about traditional fisheries management in Palau should be required reading for anyone involved in fisheries management. Prior to the imposition of Western-style national fisheries agencies in the western Pacific, community-based management was the norm. While there may be a tendency to over-glorify the way that man lived sustainably with nature in traditional communities, there is ample evidence that many communities around the world sustained much of their food intake by fishing.

I will let Johannes speak for himself. "The most important form of marine conservation used in Palau, and in many other Pacific islands, was reef and lagoon tenure. The method is so simple that its virtues went almost unnoticed by Westerners. Yet it is probably the most valuable fisheries management measure ever devised. Quite simply, the right to fish in an area is controlled and no outsiders are allowed to fish without permission."

Having exclusive access to fish resources is a prerequisite to good management. On a broader scale, the imposition of 200-mile exclusive economic zones in the late 1970s was an essential step for all countries to begin managing their fish resources. As we saw with the Chilean artisanal fisheries, it was not until the caletas could exclude others from their fishing grounds that they could manage the resources. Traditional fisheries management around the world has taken many forms and used many tools. Gear restrictions, time and area closures, and indeed permanently closed areas were all part and parcel of traditional management. It is hard to say how effective these measures were. Historical evidence certainly shows that fishing by traditional societies depleted some fish resources, but many communities were sustained for thousands of years by marine resources.

What are the characteristics of territorial fishing rights?

Territorial fishing rights, sometimes called TURFS (territorial user rights to fish) have been proposed as a major "new" tool for fisheries managers. Based on the historical experience documented by Johannes and others, and the recent experiences with the Chilean artisanal fisheries and other communities, cooperatives, or organizations that have been granted TURFS, they appear to offer a way to manage many fisheries not amenable to Western-style top-down management. The key element of TURFS is exclusive access in space, so obviously they are primarily useful for largely sedentary resources. Almost all TURFS that have been implemented are community based, although one can think of shellfish aquaculture leases as a form of TURF, where an individual or company is given the right to use a section of coast for their own activity.

TURFS offer major advantages in two areas: enforcement of illegal fishing and data collection on very small scales. Central governments will rarely have the resources to prevent illegal fishing for high-value resources like loco, abalone, and lobster. TURFS provide very strong incentives for communities and local fishermen to monitor and prevent illegal activities. Many marine invertebrates such as crabs and abalone are managed by minimum size limits, with the size limit ideally set above the size at sexual maturity. This allows for the maintenance of a minimum breeding stock. However, the same invertebrates usually have different growth rates in different habitats so that the appropriate size limit may differ from one side of an island to another, or one side of a reef to another. This level of spatial control is very difficult for a central government to identify and regulate. It is ideally suited to local control.

What are the general lessons for successful management of small-scale fisheries?

Elinor Ostrom is a political scientist who won the Nobel Prize for her studies of community-based management of natural resources. She has shown that social institutions can be an effective tool to prevent the tragedy of the commons. A 2011 study of 130 fisheries confirmed her results for fisheries. The key elements for success appear to be the need for exclusive access, social and political leadership, and cohesiveness. When communities do not have a legal framework for exclusive access, or they are not organized and united, community-based management will likely not succeed.

12

ILLEGAL FISHING

Is illegal fishing an important problem in overfishing?

On August 7, 2003, the Australian patrol vessel *Southern Supporter* spotted a vessel thought to be illegally fishing within the Australian 200-mile economic zone around Heard Island, 2,400 miles southwest of Perth in the southern Indian Ocean. The ship was the *Viarsa 1*, registered in Uruguay. She would not stop, let alone be boarded, and thus began a 3,900-mile chase across the southern ocean that had news watchers glued to their TVs for 21 days. Eventually, the *Viarsa 1* was boarded well south of South Africa, and it had 95 tons of valuable Patagonian toothfish on board. The Australian government sold the toothfish at auction for $1 million.

When we think of pirates, the great characters of the 17th century (and lately Johnny Depp as Captain Jack Sparrow) immediately come to mind. But modern piracy is alive and well on the high seas and a lot more profitable than it ever was for Blackbeard or Henry Morgan. Around the world, in international waters and in the economic zones of most nations, vessels operate illegally and annually catch fish valued at $10 to $20 billion. Perhaps as much as 30% of the total catch of fish in the world is taken illegally.

The Patagonian toothfish, also known as "Chilean sea bass" in North America, and "ròbalo" in Spain, is a large, long-lived fish found in the deep waters of the southern ocean. Its oily white flesh is highly prized in the restaurant trade, and the discovery of large stocks in and around the Antarctic Islands has generated one of the most highly publicized illegal fisheries in the last decades. The fishery began in the 1970s largely in Chile and then Argentina. By the 1990s the reported landings were 40,000 tons, worth roughly $200 million a year.

As the markets for toothfish developed, the rich resources of the far southern ocean were found, and by the mid-1990s the legal reported catch in Antarctic waters was about 12,000 tons—yet the illegal catch was estimated to be at least 32,000 tons, worth $150 million each year. Illegal fishing for toothfish has been so rampant for one simple reason: it is very profitable. Catching them is easy, they are worth a lot, and the chances of getting caught are very small. The Antarctic waters are vast and patrol boats like the *Southern Supporter* are few. Even if detected and caught, being convicted is rare. Five crew members of the *Viarsa 1* were tried in Australia for illegal fishing and were acquitted in 2005 after a hung jury. The *Viarsa 1* was ultimately scrapped.

In a classic 1968 paper, "Crime and Punishment: An Economic Approach," G. S. Becker argued that crime needs to be considered an economic activity rather than an aberrant form of social behavior. People will engage in illegal activities when it is profitable. Think of the United States during prohibition. The price of illegal liquor was high, and Al Capone and many others got very rich. Potential poachers weigh the potential benefits and the probability of being caught and the cost if they are caught. An individual, a large company, or a group of investors might make a tentative reconnaissance trip to the southern ocean to see about toothfish. If it is profitable and they believe the probability of capture is slight, they will repeat the trip and perhaps outfit another boat or two. Vessels fishing legally may be tempted to augment income with some illegal catch. Again they weigh the potential profits versus the risks. If profits are high and risks are low, some will not resist the temptation.

There is an established fisheries management system for the Antarctic toothfish stocks. *The Convention for the Conservation of Antarctic Marine Living Resources* (CCAMLR, pronounced CAM-a-lar) is an international organization whose mission is to conserve the biota of the Antarctic region. Established in 1982, with headquarters in Hobart, Tasmania, it assesses the status of resources in its jurisdiction, organizes research, and sets regulations. Also, like its sister organizations, it has neither patrol boats nor airplanes and relies on the 31 member nations to enforce regulations.

CCAMLR and its member countries rely on a range of measures to assure compliance with the catch limits set by the commission, including satellite tracking of all vessels in CCAMLR waters, mandatory port inspections of vessels returning from trips, vessel registries, and vessel marking. There is also a catch documentation scheme that documents and tracks the legal catch from the time the fish come on board until they reach retail markets.

CCAMLR estimates that their measures have reduced the illegal trade considerably and reported that for 2004–2007 just over 10% of the catch in their region was illegal. However, international conservation organizations, using trade figures, estimated that the illegal catch was slightly higher, between 14% and 23%.

While the Antarctic continent is treated as an international zone for fishery management purposes, many of the islands in the southern ocean are under national jurisdictions, including the Heard and Macquarie islands of Australia, South Georgia Island (UK), and Kerguelen and Crozet Islands (France). These countries maintain enforcement and management within their own territorial waters. The UK fishery in South Georgia was certified by the Marine Stewardship Council as well managed in 2004. The international fishery in the Ross Sea was certified by MSC in 2010, and the French fishery is applying for certification.

Is the illegal fishing of Patagonian toothfish unusual?

The world's fisheries are stretched tight between two competing traditions. There is a long historical tradition of *freedom of the seas*—in international waters you can largely do what you want when you want. Before the 200-mile economic zones and international agreements, fishing on the high seas was totally unregulated. In contrast, within national jurisdictions, fishing is one of the most heavily regulated industries. Legal fishermen are told when and where they can fish, what gear they can use, and often how much they can catch. In many places boats must carry satellite transponders so government agents can track their minute-by-minute movements. Often they are required to carry government *observers* to make sure they obey the many rules that constrain them.

The potential profits of pushing the boundaries are very high. Almost every fisherman I know has told me tales of making a big catch by some violation of the rules. Where the regulators and fishermen are antagonistic, rule breaking is an accepted way of business. If you can get away with it, do it. Becker's vision of illegal activity as an economic choice is at work every day in a fishery somewhere.

The international community calls illegal fishing IUU—Illegal, Unreported, and Unauthorized. There is some form of IUU in almost every fishery in the world, whether it is industrial-scale illegal fishing for Patagonian toothfish or the sport fisherman who keeps an undersized fish. Rule breaking is an unavoidable aspect of fishing. The most recent estimates are that about 20% of the catch in various regions around the world is illegal, with a slight decline from 1980 to 2003. Naturally these estimates are imprecise because the catch, by definition, is unreported.

How can some toothfish fisheries be certified as well managed while substantial illegal harvesting continues?

The most widely used certification, by the *Marine Stewardship Council*, applies to individual fisheries, not to species. In 2004, the toothfish fishery in the British-controlled South Georgia Islands was certified. The British demonstrated that the fishery met the standards of certification. This included demonstrating that within the British-controlled zone, illegal fishing was controlled and a harvest strategy was in place that provided for the sustainable management of the toothfish resource. The original certification was appealed by several NGOs but was upheld by a second scientific review. The fishery was recertified in 2009.

The differences in stock status and management effectiveness between fisheries for the same species is a complexity that really matters in certification or consumer information. The largest cod fishery in the world takes place in the Barents Sea, north of Norway and Russia, and was certified by the MSC in 2010. The stock is at high abundance and not considered overfished in any sense. Many other cod stocks, though, remain at low abundance. You cannot simply say that "cod" are not sustainably managed; you have to be very specific about individual cod stocks.

What methods can be used to reduce illegal fishing in international waters?

The primary tools to combat illegal fishing are (1) vessel marking, (2) vessel registry, (3) landings inspection, (4) satellite monitoring of vessels, (5) catch documentation and tracking, and (6) blacklisting offending vessels. The registry, marking, and satellite monitoring mean that vessels not registered are easily identified if detected. An airplane or ship on an enforcement survey knows exactly where every legal vessel is at any time. The landing inspections and catch documentation mean that any shipment of toothfish anywhere in the world can, in theory, be tracked back to when, where, and by which vessel it was caught. This is the same suite of tools that is commonly used for preventing IUU fishing in industrial fisheries both nationally and internationally.

13

TRAWLING IMPACTS ON ECOSYSTEMS

How do trawls and dredges work and why are they still used to catch fish?

"Enormous bottom trawl nets are dragged along the sea floor, catching all marine life and killing all habitats—they swallow and destroy everything in their path."

This is how an environmental website described what is possibly the most worrying form of fishing. Bottom trawling and dredging means dragging heavy nets along the bottom of the ocean. There are estimates that each year an area as big as the entire United States is trawled. A scientific paper likens this to clear-cutting the Amazonian rainforest once a year.

One aspect of overfishing is the impact of fishing on the marine ecosystem, and there is no better place to start than trawling. To look closer at trawling and bottom ecosystems, let us first go to the town of New Bedford, Massachusetts.

Driving through New Bedford a few years ago I was struck by the wonderful old mansions on the hillside above the waterfront, eloquent testimony to the long-gone wealth of the Yankee whalers. New Bedford was home port to most of the 19th century whaling fleet and was one of the richest towns in the country. Successful captains displayed their wealth with mansions while their anxious wives walked the daintily carved "widow's walks," high on the roof, where they could look out to sea hoping to see a familiar ship return after many years away. Now the decline of New England manufacturing has left its mark with signs of years of decay. But downtown there is new life, the waterfront is busy. Once again, New Bedford is home port to one of the most valuable fisheries in the United States, the Atlantic sea scallop. Once again, successful captains are getting wealthy and a good crewman can make as much as $100,000 a year.

Atlantic sea scallops are relatives of the clam; they live around 100 meters down and are found primarily on firm sand and gravel bottom, avoiding areas of very fine sediments like clay. They are filter feeders; they suck in seawater through their siphon and take out very small particles such as phyto-plankton, zooplankton, and eggs and larvae of other creatures including their own. They grow very fast and become sexually mature when they are three to four years old. Not totally sedentary, some tagged scallops have been known to move up to 48 km.

They are fished with dredges towed along the bottom. The dredges are heavy metal frames that scoop up the top layer of the ocean floor. A wire mesh sieves out the scallops from sand and gravel. New Bedford boats typically tow several dredges at the same time, sweeping a path along the ocean floor. The dredges are then pulled back up and the scallops are sepa-rated from any material too big for the sieve. The crew open and clean them on board. Starting at an annual catch of 3 million pounds, worth $5 million in 1973, the catch has grown 10 times—to over 30 million pounds, worth more than $200 million. It is a great success story of U.S. fisheries management from an economic and yield point of view—but those dredges are still getting dragged across the ocean floor.

Bottom trawling is very similar. The simplest and oldest kind is a beam trawl; a large beam or board holds the mouth of the net open and a footrope or "tickler" drags along the sea floor, often with small rollers attached. More common now is the "otter trawl"; its net is held open by large otter boards or doors that act like wings to spread the net. In both kinds of trawling the heavy parts of the net touch the bottom, especially the very heavy doors of an otter trawl, and dig furrows like a giant plow. The footrope itself also scrapes along the bottom. While most trawling is done on soft mud, sand, and gravel floor, some trawl gear has been designed to move over very rough sea floors that would destroy normal trawl nets. Such gear, called a "rock-hopper," has tires or wheels along the footrope to lift the net over boulders that would otherwise snag it. The development of rock-hopper gear has allowed trawl fishing to move into very sensitive habitats such as corals, raising even more environmental concerns.

There is simply no question that trawls and dredges change the ocean floor. The effects depend on the type of habitat and the amount of existing natural disturbance. Fishing gear that contacts the ocean floor is most destructive in places where there is abundant and highly structured sea life such as corals and sea fans. Trawling quickly removes most of it and leaves a very different ecosystem behind. A picture of a lush ecosystem with abundant structure before and after trawling will indeed remind you of a clear-cut forest. Such pictures carry a punch and are often used by conservation organizations to solicit donations for their anti-trawling campaigns. Almost every environmental organization opposes trawling and many are working for an outright ban or at least substantial reductions. Consumer action groups like the Monterey Bay Aquarium generally will not recommend any seafood caught by bottom trawls or dredges.

Why, then, are trawls and dredges still used?

They make money. But before outrage sets in consider this: about 20% of the world's fish catch comes from trawls. Those 20% are very important to the world food supply. Without it we will need to spread more fertilizer and pesticides on land and will have to cut down more native forests for more arable land. Everything has its price. While some of the trawled species could be caught by hook and line or with pots and traps, many others such as the Atlantic scallop can be caught only by being scraped off the sea floor.

Is trawling the ocean like clear-cutting the forest?

As in most issues of overfishing there is no simple answer. At one extreme, on soft sea floors that are cyclically disturbed by storms, trawling has little if any impact. At the other extreme, for sensitive, highly structured, and rarely disturbed habitats, the analogy to clear-cutting is apt. To really understand the impact of trawling we must look at the habitat being trawled and its natural disturbance cycle. To illustrate, here are two very well studied situations in Australia.

Keith Sainsbury, a marine biologist who worked for many years for the Australian Commonwealth Scientific and Industrial Research Organization, CSIRO (pronounced sigh-row), surveyed the tropical waters in the northwest of Australia where Taiwanese trawlers were licensed to fish for a wide range of species. He documented the differences between the trawled and untrawled areas. Areas that were trawled had a much lower diversity of sea floor life, including everything from sea fans to corals to fish, than areas that were not trawled. But once trawled areas were closed, the sea floor structure gradually recovered and the more valuable fishes came back. Traps could just as well be used to harvest the valuable species without the habitat damage of trawling. For his work Sainsbury won the prestigious Japan Prize that includes a substantial cash award and dinner with the Emperor of Japan.

At the eastern end of Australia, another CSIRO scientist, Roland Pitcher, studied the impact of trawling in a much better known area, the Great Barrier Reef, where an important prawn fishery also uses bottom trawls. The Great Barrier Reef consists of a complex network of coral reefs along the northeast coast of Australia with coral islands and subsurface reefs dotted across a generally sandy sea floor. The prawn fishery stays away from the reefs but trawls between them on the sand. Tropical cyclones with enormous wave energy that churn up the sand are frequent here. Pitcher and his colleagues set up a series of experiments to close some areas to trawling and to trawl new and untouched areas. There was very little measureable impact of trawling. A science aphorism says that "if you need statistics you didn't do the right experiment." Pitcher and his colleagues needed a lot of statistics to find any impact of trawling at all. The sandy sea floor, the currents, and the frequent tropical storms already prevented any of Sainsbury's sensitive species from living there.

What we see from these and hundreds of other studies is that the impacts of trawling depend very much on the habitat. The extreme claim that trawls catch all marine life and kill all habitats is certainly not true for most of the ocean that is trawled.

The best evidence comes from those parts of the ocean that are heavily trawled and also well studied. Three such areas are the North Sea, the northeastern United States (where the scallop fishery is located), and the Gulf of Mexico. Each one of these areas has been trawled intensively for a century. In New England, on average, every place is trawled once a year. Some habitats are trawled many times a year, others not at all. In the Gulf of Mexico, the average spot is trawled twice a year.

But after a century of industrial trawling, each of these places still produces fantastic amounts of fish on a sustainable basis, and in each one the commercially important species recover when overfishing is stopped. Haddock and cod in both the North Sea and New England are rebuilding or have rebuilt to target levels. Red snapper in the Gulf of Mexico is increasing. This could not have happened if trawling had "killed" their habitats. But it is also certain that trawling and dredging alter ecosystems (in some locations quite a bit); trawling has been shown to reduce the growth rates of some species, and some species may not recover in trawled habitats.

To truly understand the impact of bottom contact gear around the world we need to know how much of what sort of sea floor is being trawled each year. Unfortunately, we don't know the answer to this. Habitat mapping is quite incomplete and there are no summaries of what portion of the trawls are on mud, sand, gravel, hard bottom, or coral. In general, trawl fisheries avoid hard sea floors. The rich scallop beds of New England, the Mid-Atlantic coast, and the Gulf of Mexico are primarily on soft sea floors where trawling has the least impact.

Trawling is not like clear-cutting.

Most trawlers go over the same places year in and year out. Loggers cannot do that—as there are no trees left once an area has been clear-cut. Fishermen can move over the same space because they know the fish will be there again and again and they will not catch their nets on a rough sea floor. Trevor Branch from the University of Washington looked at several years' worth of records from onboard observers of every trawl for bottom fish in western Canada. Each fisherman had a set of perhaps 50 or 100 known trawl lines that were recorded on his GPS system, and each boat fished these lines on a regular basis. They certainly do not destroy one place with trawls and then move on like Amazonian loggers. Clear-cutting is not how most industrial trawl fishing works; the fisheries would have disappeared long ago if they were systematically decimating the habitats the fish need to survive.

There is a legitimate concern that trawling is expanding into new territory, particularly to deeper sensitive habitats. The United States and New Zealand preemptively closed large sections of deep water to trawling before potential fisheries had a chance to get established. There is also an international effort by NGOs to make the ban worldwide.

How long do ecosystems take to recover from trawling?

It depends greatly on the habitat, and especially on how often they get disturbed by natural events. Highly structured habitats with long-lived attached species such as soft corals may take centuries. Soft sea floors accustomed to frequent natural disturbance may return to their untrawled state in a few years.

Are there alternatives to trawling and dredging as ways to catch fish?

Many of the species that are trawled and dredged can often be caught by hook and line or in traps. In shallow waters these species can be collected by hand or spearfishing. In some cases, hook and line and traps can be as efficient as trawling and economically competitive with it. Efforts are already being made to see if some trawl fisheries can be replaced by other gears, and in places where trawl fisheries compete, to see whether catches could be re-allocated. But as of 2011, about 20% of the world's fish catch comes from trawling and for most of this catch there are simply no economically viable alternatives on the horizon.

14

MARINE PROTECTED AREAS

What are marine protected areas?

One of the crown jewels of marine ecosystems is the Great
Barrier Reef (GBR) off the northeast coast of Australia. The
reef stretches along 2,600 km of the Queensland coast and
consists of 900 islands and 2,900 reefs. It is a World Heritage
Site with some of the highest biodiversity in the world. It is
also one of the most protected marine areas in the world. The
Great Barrier Reef Marine Park Act of 1975 established most
of the GBR as a park, protected by the Great Barrier Reef
Marine Park Authority (GBRMPA). As of 2011, 33% of the
area is closed to extractive activities such as fishing; in the
other areas a blend of activities is permitted.

The four primary human activities in the GBR are
tourism, recreational fishing, commercial fishing, and
shipping. The GBRMPA relies on a technique called *marine
spatial planning or ocean zoning* to separate them to reduce
conflict. Essentially, portions of the reef are set aside for
each activity and some sections are entirely closed, even to
tourism. Marine spatial planning is increasingly advocated
as a better way to manage marine ecosystems, and the GBR
is often held up as a prime example of how spatial planning
can work both to protect marine ecosystems and to provide
for sustainable use.

Would it were this simple. Threats to the biodiversity of the GBR include (1) climate change, particularly increasing temperatures, ocean acidification, and rising sea level; (2) pollution, primarily sediment and nutrient runoff from agricultural areas on the mainland; (3) oil spills; (4) invasive species and outbreaks of coral-eating predators; (5) fishing; and (6) habitat damage from fishing gear, boat anchors, and shipping accidents. The problems associated with climate change are so profound that no actions the GBRMPA or the Australian government can take would have a significant effect. Land use policies on the mainland are also outside the direct control of the GBRMPA, but the Authority does actively engage with other agencies to try to minimize the pollution issue, and a number of agreements have been made to achieve this. The threat from oil spills generated by exploration and drilling has been eliminated by a complete ban on oil exploration or production within the GBR, and the zoning system minimizes the impacts of fishing and habitat damage.

Marine protected areas (MPAs) are areas of the ocean closed to some forms of human activity. Fishing is most commonly regulated, but oil exploration, oil drilling, seabed mining, and tourism are also potentially restricted. Establishing an MPA does not necessarily mean total protection. While some MPAs may be completely protected, there are degrees of protection associated with the general status of an MPA. Trawling and dredging that harms plants and animals attached to the bottom are most commonly restricted. Commercial fishing with different kinds of gear may also be banned, along with recreational fishing with hook and line or any combination of fishing methods. Tourism can be restricted by prohibiting anchoring or by banning human presence altogether.

The term *marine protected area* is not terribly specific, and sometimes it simply means an area with higher levels of protection than its surroundings. Perhaps more relevant to overfishing is the term *marine reserve*, which usually denotes areas closed to all forms of fishing.

What do marine protected areas protect?

The 2010 Deepwater Horizon oil spill in the Gulf of Mexico illustrates that most MPAs don't protect the ecosystem from most major threats. Oil can wash hundreds of miles. MPAs provide no protection from ocean acidification, warming seas, and sea level rise, nor do they protect from land-based pollution causing dead zones and silt from runoff, or from the major threat of exotic species, or even from illegal fishing. The current implementation of MPAs simply protect marine ecosystems from fishing. So perhaps we are collectively a bit smug when we say we have "protected" the ocean with an MPA.

How much of the world's oceans are now closed to fishing?

Closed areas have a long tradition in fisheries management, from the ancient form of traditional community-based management to Western style top-down fisheries management agencies. Any map of permitted fishing activities would look like a crazy quilt. Some areas will be closed to fishing for the protection of spawning stocks or juveniles and some to avoid by-catch; in others one type of fishing gear will be prohibited to provide an advantage for another. However, areas closed to all fishing, the true marine reserves, represent only a small portion of the world's oceans.

Several international agreements have targets to set aside 10%–20% of the oceans for MPAs, and many countries have their own specific targets for their own marine regions. Overall, as of 2007, only 1.6% of national economic zones are in MPAs and only 0.2% are in marine reserves.

Some protected areas are quite large. The Great Barrier Reef was the largest until 2000, when the United States established the Northwest Hawaiian Islands National Monument. Some countries have closed very large areas to trawling. The United States has closed more than two thirds of its 200-mile zone to bottom contact gear, although most of this is in Pacific waters too deep to be fished, and New Zealand has closed 30% of its 200-mile zone to trawling.

What is the impact of closing areas to fishing?

The difference in abundance between areas that are fished and areas protected from fishing depends greatly on two factors—how much fishing goes on outside the reserve and the size of the reserve in relation to how much the fish move around.

Studies that compare abundance inside and outside long-established reserves typically find that fish inside are two to four times more abundant.

If a reserve is small and the fish move a lot and far away, there may be no effect at all. But if the fish don't go far and fishing pressure is high outside, we may well find 5 to 10 times more abundance inside the protected area. The more an area is overfished on the outside, the greater the relative abundance of fish will be on the inside.

Not only will there be more fish but there will also be more species, or higher *biodiversity*, inside a reserve. Reserves with overfished surroundings will typically show a 30% increase in species counts, and the fishes will live to a ripe old age and trophy size—a natural consequence of not ending life early on a hook or in a net.

All well and good, but where will the fishermen go? Elsewhere, of course. And there we have a negative consequence of closed areas that can even lead to overfishing or increased by-catch outside the reserves. Such dislocation and redirecting of fishing effort can be rather disruptive to fishing communities. Longer travel means extra fuel, and extra fuel and more travel means more greenhouse gases. Longer travel also means more risk of accidents and reduced profit, which can, in the worst case, make it impossible for some types of boats to fish at all.

MPAs may also effectively "lock up" a portion of the fish stock and thereby lower the total sustainable yield. If that part of the stock stays inside the reserve, we can expect a proportional loss of sustainable yield to the outside fishery. If, however, the stocks outside the reserve are overfished, the eggs and larvae dispersing from the reserve can actually increase yield.

Do MPAs increase the abundance of fish?

In almost all cases we have established that there are more fish inside an effectively enforced MPA than outside it. But when it comes to how many more fish there are in the overall ecosystem, things get murky. We generally expect that when fishing moves outside the reserve, the added effort on the outside will decrease the abundance there. But to move from expectation to fact, we would need good data for inside and outside the reserve before and after exclusion. Alas, good data are sparse. In some cases we know that abundance did increase inside and outside a reserve after closure, but the studies lack *controls*, or data for a similar area without superimposition of an MPA. And if the establishment of an MPA just happened to coincide with good environmental conditions for everyone, we would certainly expect abundance to increase outside the reserve as well. In other cases, we know that abundance increased inside the reserves but declined outside.

In general, ecological theory expects and predicts that if overfishing is a major problem, establishing an MPA will result in more fish in the system overall. This is because eggs and larvae drift out of the reserve and reseed the adjacent overfished areas and thereby increase overall abundance.

Can MPAs solve some of the problems of overfishing?

Yes, when fisheries are overexploited and it is impossible to regulate either fishing effort or catch, MPAs are often effective at maintaining populations of fish within the reserves—but only if the local fishing community respects the area as protected. Both traditional and modern managers of small-scale fisheries around the world now use protected areas as one of their management tools.

In areas where a fisheries management system that prevents yield overfishing is already in place, MPAs must be seen primarily as natural parks where more pristine levels of abundance and community structure are preserved and where the ecosystem is protected from overfishing. Still, MPAs will probably have no part in preventing economic or yield overfishing.

And this is precisely why MPAs are so controversial in developed countries with established fisheries management systems. Recreational and commercial fishermen believe that heavy regulations to prevent overfishing already exist and that there is really no need for MPAs to add to their burden.

How much of the ocean should be set aside as protected from fishing?

It all depends on the objectives—what should the protected areas accomplish and for whose benefit? There are already international targets to lock up between 10% and 30% of national economic zones under various levels of protection, but most countries are very far from those targets. In countries with well-functioning systems to prevent overfishing, reserves are primarily intended to provide protection for representative habitats and their associated biodiversity, much like terrestrial national parks. For comparison, about 10% of the United States land area is located in parks. Here, as with most aspects of ecosystem-based management, the answer depends on social choices, not scientific analysis.

15

ECOSYSTEM IMPACTS
OF FISHING

How does overfishing affect ecosystems?

Early explorers' accounts are full of wonder at the natural wealth of the New World and astonishment at the size of fish. Writing in 1615 about John Cabot's voyage off Newfoundland, Peter Martyr said, "in the sea adjacent [he] found so great a quantity of great fish that at times they even stayed the passage of his ships."

Fishing changes ecosystems. The more intense the fishing, the larger the effect. Heavily fished systems may well be totally transformed.

Changes come about in many ways. The direct removal of individual fish, indirect impacts through removing predators or prey, and physical impacts of fishing gear all play their part. Older fish are almost always the first to be targeted, and as fishing pressure increases, size and abundance go down. Even in systems that are sustainably managed for long-term maximum yield, abundance is expected to be only 20% to 50% of what it would be in the absence of fishing. Overfished systems will typically be even more transformed and overall abundance can be as low as 10% of pre-fishing numbers.

When fishing selectively removes certain species, the balance of predation and competition changes. Remove a predator, and its prey and competitors will multiply, whereas the predators of the now absent fish will find themselves short on food and their numbers will decrease. Therefore, we expect there to be fewer marine birds and mammals in a fished ecosystem since much of the energy that makes its way up the food chain will be diverted for human consumption and is therefore no longer available to them.

This demonstrates that there will always be winners and losers from fishing.

If we close areas to fishing we typically find that the species we cease to fish, the *target species*, will increase in abundance. At the same time, many species we did not fish before, *non-target species*, will decline. Non-selective gear like trawls or gill nets will catch most species in their path, and the popular literature is full of accounts of how such practices "destroy" ecosystems. On the whole, this is not an unrealistic description of how fishing has transformed many places.

However, it is important to emphasize that sustainable fishing will also result in lower abundance and average size, and that this is simply unavoidable if we are to depend on the ocean for food.

Of course, as soon as we move from sustainable fishing to overfishing, the impacts become greater. Environmental impacts exist on a continuous gradient. Very little fishing produces very little food and has very little impact on the ecosystem. Sustainable fishing produces a lot of food and changes the ecosystem considerably. Severe overfishing produces little food and completely transforms the ecosystem.

Fishing gear also changes ecosystems. As discussed in chapter 13, fishing gear that is drawn along the bottom (bottom trawls and dredges) will take out much of the biological structure that is rooted to the sea floor such as sea fans, corals, and a host of other species. Even traps and bottom line fishing can modify the structure of a bottom community if they are dragged along the sea floor.

When debating appropriate levels of fishing there is now nearly universal agreement that we should reduce fishing pressure to a level that will give us the most food over the long term, but there is no agreement at all about whether we should fish less than that.

There are severe "protectionists" like Sylvia Earle, the well-known ocean explorer, who feel that we should leave the oceans alone and not fish at all. In direct opposition are most governmental fishing policies that seek to manage fisheries at levels that give us the most food.

Remember that between not fishing at all and fishing for the best long-term food production, there is a continuum of fishing pressure that is sustainable but has different consequences. Countries that consider food production most important will likely choose to fish harder than countries more interested in best profit, whereas countries that value pristine ecosystems more highly will fish even less.

Are coral reefs particularly sensitive to fishing?

Coral reefs do appear to be particularly sensitive, especially those reefs close to human settlements, as they are subjected to long and often intense fishing pressure. Dynamite fishing can completely destroy the physical structure of a reef, as can bottom trawling on soft corals in deeper water.

The complex interactions between fish, algae, sea urchins, and corals are ecologically very interesting. Removing key herbivorous fish can cause some algae to increase and smother corals. Fishing out key predators, on the other hand, can release sea urchins to multiply and chomp on too many of the crustose coralline algae that often make up the base of the reef's food chain.

Studies comparing the abundance of fish across a range of coral reefs in the Pacific found that the abundance of fish around heavily populated islands was about 25% of the abundance around unoccupied ones. Large predatory fish did particularly poorly near lots of people.

One of the biggest threats to coral reefs is coral bleaching, usually associated with particularly warm water years. Bleaching happens when small microorganisms known as *zooxanthellae*, which live inside the corals, are either expelled by the corals, die, or lose their pigments—and the corals turn white. The zooxanthellae are needed for the photosynthesis that helps maintain the coral. If bleaching continues and the zooxanthellae do not return, the coral itself dies. There is good evidence that corals are more prone to bleaching when the fish community associated with the reef has been heavily fished.

What is a trophic cascade?

In many ecosystems, predators and prey are tightly linked. For instance, along much of the Alaskan coast, sea otters act as intense predators of sea urchins. In turn, sea urchins feed heavily on kelp forests. When hunters began to target sea otters for their fur, sea urchin populations exploded and decimated kelp forests in many places.

When, as happened in Alaska, a predator high on the food chain disappears and causes a cascade of ecological changes downward through the food web, it is called a *trophic cascade*.

Steve Carpenter and Jim Kitchell, two ecologists at the University of Wisconsin, gave a dramatic experimental demonstration of how trophic cascades could work. They performed their experiment with piscivorous fish, which eat other fish; herbivorous fish, which eat zooplankton; the zooplankton; and phytoplankton. Simulating the effect of sport fishing, Carpenter and Kitchell removed most of the piscivorous fish. They found that the prey of the piscivorous fish, the fish that eat zooplankton, then increased in abundance—

promptly sending the zooplankton into a decline and eventually increasing the phytoplankton. Fishing at the top of the food chain in this case causes a wave of changes all the way down to its lowest point.

It is not clear how common trophic cascades are in marine ecosystems and how much responsibility fishing bears for their occurrence. The removal of herbivorous fishes from coral reefs, which, as we saw, allows algae to increase and smothers the corals, is one example, and there are certainly many more examples where there is only one link in the cascade. Yet two factors mediate against the phenomenon by which fishing causes a multitude of trophic cascades. First, fishing often removes a broad range of species across the eco-system. It is a rare system where only the top predators are taken, instead we tend to catch high and low trophic-level species. Second, marine ecosystems are characterized by many species with highly mixed diets. Most fishes are flexible in what they eat and will readily switch to other species. The simple A eats B eats C eats D phenomenon is the exception rather than the rule.

Do forage fish need special protection?

Forage fish describes the species that constitute the main prey of most of the fish-eating birds, mammals, and other fish in a marine ecosystem. The most common forage fish are sardines, herrings, anchovy, capelin, sprats, and shad. They are typically filter feeders and live primarily on the zooplankton. They are among the most abundant fishes in the sea and often form schools, which makes them particularly easy to catch. Most of the high-volume fisheries of the world, such as the anchoveta fishery of Peru; the herring fisheries of Europe; the sardine fisheries of Japan, California, and South Africa; and the menhaden fishery of the southeastern United States, catch forage fish.

Clearly, then, fishing down the forage fish means that there is less food for everyone higher up on the food chain, something to remember when evaluating how hard to fish. Historically, managers in the Western countries have considered the sustainable yield of each species individually. For example, calculations of the potential maximum sustainable yield of sardines in California does not include consideration of any impacts on either the forage fish that eat sardines or the birds and mammals that eat the forage fish. But fisheries managers now often do include allowances for species higher on the food chain.

As soon as such factors are taken into consideration, it seems clear that it would be advisable to fish the forage fish less hard than single-species management suggests. This also suggests that there are other species of fish and marine birds and mammals that might be commercially or recreationally important. The actual calculation of these impacts is often difficult and imprecise, but consensus is emerging that we should fish forage fish less hard than we would for sustainable yield.

A related and highly controversial topic is the potential fishing of krill, large invertebrates much like tiny shrimp that form the basis of the food chain, especially in the Antarctic. Krill make up the largest base of the forage fish microorganism diet, and they are also the primary prey of the large baleen whales. They are wonderfully abundant, and it is estimated that the sustainable yield of krill might be as much as the entire harvest of all other animals from the ocean. The problem is that if we fish krill, there will be less food for whales and other Antarctic species.

What is by-catch and how important is it?

By-catch is a term applied to the unintended or undesired catch by fishing gear. Of particular concern is by-catch of birds, mammals, sharks, and sea turtles that may be endangered. But mostly by-catch consists of non-target species that are thrown away because they have no commercial value, or species that need protection because they are overfished.

Discards are of primary concern because of the waste. It was estimated that in the mid-1990s about 25% of the fish caught (which works out to 27 million tons) were thrown overboard. This number is thought to have declined considerably, largely because there are now markets for much of what was formerly thrown away; nevertheless, we will never know how much went overboard because discards are, by definition, not landed and sold.

The amount of by-catch and discards is highly variable among fisheries. Shrimp and prawn trawls are among the worst, with an average of more than five tons of fish discarded for every ton of shrimp or prawns landed. At the other extreme are fisheries for pelagic fish caught in single-species schools or the fisheries for pollock in Alaska, where the by-catch is often much less than 1% of the catch.

There are three major approaches to reducing by-catch and discard. Technological solutions have been developed that modify how fishing is done. Perhaps best known is the attempt to prevent dolphins from being caught in the tuna purse seine fisheries of the eastern Tropical Pacific. Now the fishing boats have a special procedure called *back down* where a portion of the net is lowered to allow the dolphins to swim out of the net before it is brought on board. Turtle excluder devices prevent turtles from being caught in the end of the shrimp trawls. Longline fishing has been changed in many places to avoid sea birds diving on the baited hooks and getting caught. Another approach is temporary or permanent closure of areas with high by-catch. Finally, by-catch limits can be imposed on individual vessels or entire fleets, and these limits are very strong incentives for fishermen to find ways to limit catching non-target species.

How does ecosystem-based management differ from single-species management?

Ecosystem-based management or the ecosystem approach to fisheries management recognizes that species in marine ecosystems are interconnected. It further recognizes that both the fishermen and the management system used must be considered in the fisheries management process. The fisheries management approaches that emerged in Western societies in the 20th century tended to be single-species approaches: look at the fish stock, determine how much fishing pressure would achieve maximum sustained yield, and regulate fishing effort or catch to achieve that level of fishing pressure. The single-species approach has many deficiencies: It does not consider by-catch; impacts on predators, prey, and competitors; or impacts of fishing gear. Nor does single-species management explicitly consider the fishing fleets or the management system and how it interacts with the species-by-species regulatory process. For instance, you might increase fishing pressure on one species when another species declines to stabilize total catch.

Ecosystem-based management has been formally adopted by many fisheries management agencies around the world, but the way it is implemented is highly variable. Most management organizations have explicit approaches to reducing by-catch and protecting vulnerable species, often through gear, time, and area restrictions on fishing. Increasingly, the target harvest rates are lowered below single-species maximum sustained yield, and target biomass levels are set higher than would produce maximum sustainable yield (MSY). Both effectively reduce overall impacts of fishing pressure. Many organizations are restricting the use of bottom trawls in sensitive habitats or setting aside large areas as no-trawl zones.

Conceptually, ecosystem-based management is a holistic approach, yet a major problem is that it is, by definition, multi-objective—we want more marine birds and mammals, but also more fish to eat and more jobs. It is not possible to maximize all these objectives simultaneously, and thus far legislators have shied away from specifying just where on this set of continuous trade-offs society wants to be. The result has been somewhat piecemeal implementation where the give and take of the management system results in quite variable outcomes.

A further problem is that ecosystem-based management means different things to different stakeholders. To preservationists, it means dramatically reducing fishing pressure and closing large areas of the ocean to fishing. To fishing communities, it means considering the sustainability of their communities as integral to the goals of fisheries management. As a result, in the case of a mixed fishery with some depleted stocks, the preservationists would opt for near total closures on fishing whereas fishing communities would protect their economic basis and decide on a much more gradual rebuilding program, even one that might see some species continually overfished in order to maximize the sustainable yield from the entire ecosystem. To some, ecosystem-based management to maximize food production would deliberately cull fish-eating marine mammals. Other stakeholders who place high value on marine mammals would see such actions as anathema.

What is the precautionary approach to fisheries management?

The precautionary approach to fisheries management evolved from the *precautionary principle* that actions should not be allowed until it is known that they will not be harmful to the environment. In fisheries management this approach emerged largely as a backlash to the fact that fisheries regulations were often not implemented until there were clear signs of overfishing.

The precautionary principle is a very conservative concept. No fishing would be allowed until the ecosystem impacts and the level of sustainable fishing are known. Since we primarily learn about the impacts of fishing by actually fishing, the logical conclusion of the precautionary principle is that no fishing can be allowed except in highly controlled experimental settings.

Historically, the burden of proof had been placed on those arguing for fishing restrictions, whereas the precautionary principle places the burden of proof on those proposing to fish.

The precautionary approach tries to balance the potential benefits of an action (usually allowing specific levels of harvest) against the potential risks of that action. The Food and Agriculture Organization (FAO) of the United Nations in their report on the precautionary approach to fishing (see Further Reading) dealt specifically with that burden of proof: "The standard of proof to be used in decisions regarding authorization of fishing activities should be commensurate with the potential risk to the resource, while also taking into account the expected benefits of the activities."

The specific elements of the precautionary approach to fisheries management defined by the FAO report are "(1) the consideration of the needs of future generations and avoidance of changes that are not potentially reversible; (2) prior identification of undesirable outcomes and of measures that will avoid them or correct them promptly; (3) any necessary corrective measure must be initiated without delay...; (4) that, where the likely impact of resource use is uncertain, priority should be given to conserving the productive capacity of the resource; (5) that harvesting and processing capacity should be commensurate with estimated sustainable levels of resource...; (6) all fishing activities must have prior management authorization and be subject to periodic review; (7) an established legal and institutional framework for fisheries management, within which management plans that implement the above points are instituted for each fishery; and (8) appropriate placement of burden of proof by adhering to the requirements above."

How many marine fish species are threatened with extinction?

The International Union for the Conservation of Nature (IUCN) is the most authoritative source of classification of the threat of extinction. It has established a set of criteria that put a species into categories such as extinct, critically endangered, endangered, vulnerable, near threatened, least concern, and data deficient. Extinct means the last living individual has died. Critically endangered means there is a 50% probability of being extinct in 10 years, endangered is a 20% probability of being extinct, and vulnerable is a 10% probability in 100 years. Through a series of specialist groups it has evaluated the threats to many but not all species. By now it has evaluated all marine birds, mammals, reptiles, and sharks. As of 2008 about 20% of sharks, 30% of corals, marine birds, and marine mammals, and 90% of marine turtles were considered critically endangered, endangered, or vulnerable. For sharks and marine mammals, more than 30% of species were considered data deficient and could not be assessed.

The only bony fish species completely evaluated was groupers, a tropical fish often found associated with coral reefs and intensely fished in many places. About 15% of groupers were found to be in one of the categories indicating that they are threatened with extinction. Another project randomly sampled species of all taxa and then evaluated them against the IUCN criteria. Slightly more than 10% of bony fish (excluding sharks and rays) would meet the criteria for being threatened with extinction.

There has been much publicity about the threat of extinction of orange roughy and Atlantic bluefin tuna. While both have been heavily fished and are at much lower abundance than 100 years ago, there are hundreds of thousands of Atlantic bluefin tuna and hundreds of millions of orange roughy. Any comparison to terrestrial species that number in the hundreds or thousands is a gross exaggeration. The threat to these species and other commercially important species is continued overexploitation. In the case of orange roughy, large areas of the ocean in both New Zealand and Australia where orange roughy live are closed to fishing, and so extinction from fishing is impossible. The future of Atlantic bluefin tuna depends on whether fishing pressure can be reduced.

16

THE STATUS OF OVERFISHING

Are the world's stocks overfished?

The most authoritative assessment of the status of commercial fish stocks comes from FAO, the Food and Agriculture Organization of the United Nations. Every two years FAO publishes a report that summarizes the status of commercially important fish stocks. FAO estimated that in 2008, 32% of fish stocks worldwide were overfished, depleted, or recovering.

This is an estimate of stocks subject to yield overfishing. If we consider economic overfishing or ecosystem overfishing the percentages are obviously much higher.

In 2009, I participated in an initiative to establish a database on abundance trends in fish stocks based on assessments by management agencies. This database now contains over 300 of the most important fish stocks in the world. It is by necessity heavily biased toward Europe and North America, where most fish stocks are assessed. The rapidly developing fisheries in Asia are largely absent due to a lack of public stock assessments. However, our database does represent almost all the stocks that have caused so much concern about overfishing. According to FAO's measures of overfishing, the regions covered in the database are more overfished than the rest of the world.

We found that most of the developed world had been subject to wide-scale overfishing in the 1980s and 1990s, and that at present about two thirds of stocks are below the abundance level that would produce maximum sustainable yield (MSY). Roughly one third of all stocks can be classified as overfished—their abundance is low enough that their sustainable production is significantly reduced. Thus we came to the same conclusion as FAO.

Only Alaska and New Zealand have not been overfishing any significant fraction of their stocks. In every other fishery for which we have data, overfishing has been part of its history.

More encouraging, we found that fishing pressure had been reduced significantly almost everywhere. When we looked at the data, we found that by the mid-2000s the fishing pressure in two thirds of the stocks was too low to produce maximum sustained yield. Only about 15% of stocks were fished hard enough to produce significant decreases in long-term yields. The good news is that in the areas covered by this database the threat to food security is reasonably small.

From an economic or ecosystem perspective large portions of stocks are still overfished. For one third of stocks, fishing pressure is high enough that they will on average stay below the traditional target of abundance, the level that would produce MSY. Over 60% of stocks are still being fished harder than necessary to bring maximum long-term economic returns. While the picture is very encouraging from the perspective of food security, there is much room for improvement from the point of view of economics or ecosystems.

At present we do not have data on historical fishing pressure for much of the world, particularly Asia and Africa. This makes it difficult to tell if there is overfishing on these continents, but based on ever growing catches from these regions we have no reason to believe there has been any recent lowering of fishing pressure. Nevertheless, the FAO report on status of fisheries in 2005 showed that the proportion of stocks overfished or depleted in Asia and Africa was much less than in Europe and North America. So what is widely suspected is that effort continues to grow in Asian and African fisheries and overfishing will continue to increase there since there are not the same legal and institutional structures in place that have reduced fishing pressure in developed countries.

What characterizes countries that have managed their fisheries well and those that have not?

What do we mean by well managed? To some, well managed means very little fishing and largely intact ecosystems. To others, well-managed fisheries should produce near maximum economic value to an individual nation or the world. Others might consider well managed to ensure food security, and then again it might mean maintenance of traditional fishing communities and employment.

As concerns the management of yield overfishing, the United States, New Zealand, Norway, and Iceland stand out. The United States in particular is the only country that has formally defined overfishing and has strict laws that require actions to be taken with violators. In January 2011, Steve Murawski from the University of South Florida and former chief scientist for the U.S. national fisheries management agency announced that overfishing in U.S. federally managed fisheries had ended. No other country can make that claim. In New Zealand, loss of food production from overfishing has never been significant, and it is a minor issue in Iceland and Norway.

Let us now shift from yield to economics. This is where New Zealand, Iceland, and Norway excel. In all three countries the capacity of the fishing fleet is matched to the capacity of their fish resources—no cases of too many boats chasing too few fish—and there are few if any subsidies. Iceland is particularly interesting here since before the disaster of the banking bubble in the late 2000s, it already had a very high standard of living based solely on fishing. It had also done an outstanding job of maximizing the potential wealth from the oceans. By contrast, in most other countries of the world, fishing has been a net drain on the national economies, with subsidies often almost as high as the value of fish produced. It is much harder to find countries with particularly good environmental records. The United States has been a prominent leader in declaring large portions of its economic zones as parks and protected areas. At present, probably the best indicator of environmental impacts of fishing is the overall level of fishing pressure. Countries that stand out due to their successes in reducing yield overfishing should be recognized for equally lowering their impacts on ecosystems.

It is difficult to identify countries that have been able to maintain their traditional fishing communities since there are no large-scale databases available.

The key similarities among Iceland, Norway, and New Zealand are their size and their ability to end the race for fish. They have done so well primarily because they have matched their fleet to their resources. There are no incentives for fishermen to beat the other boats to the fishing grounds, and so they have avoided excess fishing and economic loss.

When it comes to successful fisheries management, size definitely matters. New Zealand, Iceland, and Norway are small countries with relatively uncomplicated political systems. As a general rule, good fisheries management flourishes when there is a straightforward management system with only a few significant political powers and stakeholders. This gives small countries a significant advantage over the European Union (EU), which has to balance the needs of multiple countries with competing agendas. International fisheries management organizations attempting to make sense of high seas fisheries like tuna are also hampered by the necessity of having consensus from many countries.

In Europe there is a move to give fisheries decision making back to the regions within the EU. This would stop Italy and

Spain from meddling in Baltic fishing affairs, for example. This would be a very positive step if it indeed reduces the number of politically powerful groups involved in decision making.

In my opinion, the keys to successful fisheries are (1) exclusive access, (2) well-defined objectives, and (3) governance structures within the political entity.

How important are subsidies in the current problem with fisheries?

Subsidies can exist in the form of lower fuel costs; low-interest loans for vessel construction; government-funded fishing access agreements with other countries; government-financed buy-back schemes for vessel reduction; government-financed technical assistance; and taxpayer-funded data collection, research, and management. They can accomplish one of two goals: They can encourage overcapitalization and increasing effort or they can encourage better management. It is estimated that in 2000, $10 billion was spent worldwide to subsidize overfishing; a little over half of that went for fuel and the remainder for vessel construction and government-funded access agreements.

Subsidies on that scale are a major threat to the social, economic, and ecological sustainability of marine resources because all they really do is encourage excess fishing capacity and fishing pressure, which leads, in turn, to environmental, economic, and yield overfishing.

Is consumer action and certification important in stopping overfishing?

In North America and Europe it is already common to see people studying various cards produced by groups like the Monterey Bay Aquarium and Greenpeace that can be found on their websites advising them what species they should and should not eat before they buy fish. I suspect that not a lot of people in Asia, South America, or Africa carry these cards. Globally, consumer action has so far not played a role, but it is certainly emerging as a force in North America and Europe. The influence on large retailers is of particular importance. When companies with such clout as Wal-Mart in the United States or Tesco in the UK announce they will sell only certified seafood, the fishing industry takes notice.

In 2007, the Marine Stewardship Council (MSC) certified 7% of the world's fish catch used for human consumption as "sustainable." The number of stocks now being evaluated or already certified is considerably higher, and we can expect that by 2012 many large retailers will stock only fish that have been certified by the MSC. There is also considerable evidence that many fisheries are improving their management systems by documenting ecosystem impacts and by-catch to achieve certification, which definitely has ecological ramifications. However, we must remember that the seafood trade is global, and at present most of the world cannot afford to join any consumer action as discerning as the ones in North America and Europe.

How do the environmental costs of fishing compare to those of livestock?

First, foremost, and always—there is no free lunch.

We have heard much about the environmental consequences of overfishing, but let us not forget that even sustainable fishing affects the environment. In the best-managed fisheries, abundance is necessarily lower and the ecosystem different from its pristine days. Fishing also uses up other resources, particularly fuel, the source of many greenhouse gases. Best estimates are that modern fisheries use over 10 times more energy from hydrocarbons than they produce in food. Eating fish does have real environmental costs.

And so does every other form of food production. The salient question is, how high the price? Livestock production needs vast amounts of fresh water and antibiotics to raise the animals and great quantities of fertilizer and pesticides to feed them, whereas marine fisheries require very little, if any, of those. Marine fisheries have a lower carbon footprint than beef, dairy cows, or lamb because these animals generate large amounts of greenhouse gases when they digest their food. If you are concerned about carbon footprint, clean water use, pollution, or chemicals, eating fish is the environmentally friendly choice.

Biodiversity has been the main environmental concern with fishing. And here lie the fundamental differences between sustainable fishing and agriculture.

Sustainable fishing reduces the abundance of fish to between 20% and 50% of what it was in pre-fishing times. When sustainable fishing is practiced, the primary food sources—phytoplankton and other photosynthetic microorganisms—are not harvested, and the secondary food sources—zooplankton and krill—are rarely harvested. When we convert pristine land to agriculture, we rip out or plow under the native vegetation and replace it with exotic species. Truly, there is no comparison between the overall environmental impact of fishing and agriculture. Fishing has a much lighter touch. Moreover, the ecosystem in a managed fishery looks a lot more like the natural ecosystem did and maintains more of the native flora and fauna than areas converted to agriculture, like the great plains of the United States or the vineyards of Europe.

The question is not whether we should eat fish or meat. They are both important parts of food security, and we can certainly do better in reducing the environmental impacts of both. But it is good to keep in mind that the standards we have set for maintaining biodiversity in fisheries management by groups advising consumers are much higher than the standards for agriculture. I recently heard the story of a well-known environmentalist refusing to eat lobster at a banquet and insisting on a steak. I doubt that she had thought clearly about the environmental consequences of that choice.

Should we all become vegetarians?

The choice of what to eat must be made by each individual. There is no doubt that the environmental footprint of a vegetarian is lighter than that of a meat eater.

My wife used to be a vegetable farmer on Whidbey Island, north of Seattle, Washington. On five acres she raised 120 varieties of organic vegetables and sold them to restaurants, at farmers markets, and by subscription to 90 families. Her farming operation was a model of small-scale, locally grown, and organic food production. Absolutely wonderful, right?

In 1850 the five acres were covered by temperate rain forest. My wife's five acres of organic farm, the ideal of modern environmentally conscientious food production, had come at the cost of five acres of native habitat. I venture to say that not one shred of the native plants that were there in 1850 remained after spring plowing. The loss of biodiversity was total. By comparison, the oceans off the Washington coast, while different from what they were in 1850, are remarkably similar. The relative abundance of species has changed, but you will find the same species and the same diversity of habitats.

Even the costs of a vegetarian diet are considerable, and it is not clear that a vegetarian diet is necessarily one of lower environmental impact than a diet that includes fish. The choice is, of course, yours.

What is needed to stop overfishing?

When the group that Boris Worm and I organized finished its work, we found that management agencies can solve the problem of yield overfishing with the range of tools that are already commonly used by Western fisheries agencies: total catch restrictions, gear limitations, closed areas, and effort limitation. This kind of top-down management can work when governments are well funded and the fishing fleets respect the regulations, but they will do so only when they believe that it is in their interest. When the management process is considered legitimate by the fishermen and the government has the power to enforce regulations, the Western style of management can work.

As we saw in the case of Pacific halibut, solutions to the problem of yield overfishing do not necessarily apply to economic overfishing. Eliminating the competition among fishermen to catch the fish before anyone else does will go a long way, however. We can get rid of the race for fish by allowing exclusive access to individuals or groups of fishermen who use a government or social mechanism to allocate shares of the catch among themselves. There will always be friction with those who are excluded from access and those who perceive their share to be unfair.

Small-scale fisheries require different solutions. Central governments rarely have the resources to manage a great number of small fisheries. Success almost always involves devolution of power to some form of co-management, with local communities playing a major or exclusive role in data collection, management, and enforcement. Again, exclusive access appears to be a requirement for success.

Whether in industrial or small-scale fisheries, elimination of subsidies for construction and vessel operation is a first step to building biological and economic sustainability. Fisheries should be a source of great wealth to all coastal countries as they are already in Iceland, Norway, and New Zealand. It is truly sad to see so many countries squandering the potential wealth of their fisheries through excess capacity and overharvesting.

FURTHER READING

General Readings

Food and Agriculture Organization (FAO). *State of the World's Fisheries and Aquaculture 2010*. Rome: FAO, 2010. http://www.fao.org/docrep/013/i1820e/i1820e00.htm

Grafton, R. Q., R. Hilborn, D. Squires, M. Tait, and M. Williams. *Handbook of Marine Fisheries Conservation and Management*. Oxford: Oxford University Press, 2010.

Haddon, M. *Modeling and Quantitative Methods in Fisheries*. London: Chapman and Hall, 2001.

Hilborn, R., T. A. Branch, B. Ernst, A. Magnusson, C. V. Minte-Vera, M. D. Scheuerell, and J. L. Valero. "State of the World's Fisheries." *Annual Review of Environment and Resources* 28 (2003): 359–99.

Worm, B., R. Hilborn, J. K. Baum, T. A. Branch, J. S. Collie, C. Costello, M. J. Fogarty, E. A. Fulton, J. A. Hutchings, S. Jennings, O. P. Jensen, H. K. Lotze, P. M. Mace, T. R. McClanahan, C. Minto, S. R. Palumbi, A. Parma, D. Ricard, A. A. Rosenberg, R. Watson, and D. Zeller. "Rebuilding Global Fisheries." *Science* 325 (2009): 578–85.

1. Overfishing

Finlayson, A. *Fishing for Truth: A Sociological Analysis of Northern Cod Stock Assessments from 1977–1990*, Vol. 52. St. Johns, Newfoundland, Canada: Institute of Social and Economic Research, Memorial University of Newfoundland, 1994.

Harris, L. *Independent Review of the State of the Northern Cod Stock*. Ottawa, Ontario, Canada: Ministry of Supply and Services, 1990.

Hilborn, R., and E. Litsinger. "Cause of Decline and Potential for Recovery of Atlantic Cod Populations." *Open Fish Science Journal* 2 (2009): 32–38.

Hutchings, J. A., and R. A. Myers. "What Can Be Learned from the Collapse of a Renewable Resource? Atlantic Cod, *Gadus morhua*, of Newfoundland and Labrador." *Canadian Journal of Fisheries and Aquatic Sciences* 51, no. 9 (1994): 2126–46.

Rice, J. C. "Every Which Way but Up: The Sad Story of Atlantic Groundfish, Featuring Northern Cod and North Sea Cod." *Bulletin of Marine Science* 78, no. 3 (2006): 429–65.

Rothschild, B. "Coherence of Atlantic Cod Stock Dynamics in the Northwest Atlantic Ocean." *Transactions of the American Fisheries Society* 136 (2007): 858–74.

2. Historical Overfishing

Best, P. B. "Recovery Rates in Whale Stocks that Have Been Protected from Commercial Whaling for at Least 20 Years." *Report of the International Whaling Commission* 40 (1990): 129–30.

Bockstoce, John. *Whales, Ice, and Men: The History of Whaling in the Western Arctic.* Seattle: University of Washington Press, 1986.

Carwardine, M. *Whales, Dolphins and Porpoises.* London: Dorling Kindersley, 2000.

Stoett, Peter J. *The International Politics of Whaling.* Vancouver: UBC Press, 1997.

3. Recovery of Fisheries

DiBendetto, David. *On The Run, an Angler's Journey Down the Striper Coast.* New York: William Morrow, 2003.

Greenberg, P. *Four Fish, the Future of the Last Wild Food.* New York: Penguin Press, 2010.

Richards, R. A., and P. J. Rago. "A Case History of Effective Fishery Management: Chesapeake Bay Striped Bass." *North American Journal of Fisheries Management* 19 (1999): 356–75.

4. Modern Industrial Fisheries Management

Bering sea pollock species profile. http://www.fakr.noaa.gov/npfmc/sci_papers/Species_Profiles2011.pdf.

Bering sea groundfish management plan. http://www.fakr.noaa.gov/npfmc/fmp/bsai/BSAI.pdf.

Greenpeace's comments on the Alaska pollock fishery. www.greenpeace.org/usa/en/news-and-blogs/news/alaska-pollock-fishery-on-the/downloaded 5 January 2010

5. Economic Overfishing

Casey, Keith E., C. M. Dewees, B. R. Turris, and J. E. Wilen. "The Effects of Individual Vessel Quotas in the British Columbia Halibut Fishery." *Marine Resource Economics* 10, no. 3 (1995): 211–30.

Clark, W. G., and S. R. Hare. "Effects of Climate and Stock Size on Recruitment and Growth of Pacific Halibut." *North American Journal of Fisheries Management* 22 (2002): 852–62.

Clark, W., and S. Hare. *Assessment of the Pacific Halibut Stock at the End of 2002.* Report of the Assessment and Research Activities. Seattle: International Pacific Halibut Commission, 2002.

Deacon, Robert T., Dominic P. Parker, and Christopher Costello. "Improving Efficiency by Assigning Harvest Rights to Fishery Cooperatives: Evidence from the Chignik Salmon Co-op." *Arizona Law Review* 50 (2008): 479–509.

Gordon, H. S. "Economic Theory of a Common Property Resources: The Fishery." *Journal of Political Economy* 62 (1954): 124–42.

Gutierrez, N. L., R. Hilborn, and Omar Defeo. "Leadership, Social Capital and Incentives Promote Successful Fisheries." *Nature* 440 (2011): 386–89.

Hardin, G. "The Tragedy of the Commons: The Population Problem Has No Technical Solution; It Requires a Fundamental Extension in Morality." *Science* 162 (1968): 1243–48.

Ostrom, E. *Governing the Commons: The Evolution of Institutions for Collective Action.* Cambridge: Cambridge University Press, 1990.

6. Climate and Fisheries

Brander, K. M. "Global Fish Production and Climate Change." *Proceedings of the National Academy of Sciences, USA* 104, no. 50 (December 2007): 19709–14.

Cushing, D. *Climate and Fisheries.* London: Academic Press, 1982.

Doney, S. C. "The Dangers of Ocean Acidification." *Scientific American* 294, no. 3 (2006): 58–65.

Food and Agriculture Organization (FAO). "Climate Change Implications for Fisheries and Aquaculture. Overview of Current Scientific Knowledge." Fisheries and Aquaculture Technical Paper 530. Rome: FAO, 2009.

Lluch-Belda, D., R. J. M. Crawford, T. Kawasaki, A. D. MacCall, R. H. Parrish, R. A. Schwartzlose, and P. E. Smith. "World-wide Fluctuations of Sardine and Anchovy Stocks: The Regime Problem." *South African Journal of Marine Science* 8 (1989): 195–205.

Soutar, A., and J. D. Isaacs. "Abundance of Pelagic Fish during the 19th and 20th Centuries as Recorded in Anaerobic Sediment off the Californias." *Fishery Bulletin* 72, no. 2 (1974): 257–74.

7. Mixed Fisheries

Daan, N. "Changes in Cod Stocks and Cod Fisheries in the North Sea." *Rapports et Procés-verbaux des Réunions du Conseil International pour l'Exploration de la Mer* 172 (1978): 39–57.

Poulsen, R. T. 2005. *An Environmental History of North Sea Ling and Cod Fisheries 1840–1914.* Esbjerg, Denmark: Syddansk University.

Rijnsdorp, A. D., P. I. vanLeeuwen, N. Daan, and H. J. L. Heessen. "Changes in Abundance of Demersal Fish Species in the North Sea between 1906–1909 and 1990–1995." *ICES Journal of Marine Science* 53 (1996): 1054–62.

Rogers, S., and J. R. Ellis. "Changes in the Demersal Fish Assemblages of British Coastal Waters during the 20th Century." *ICES Journal of Marine Science* 57 (2000): 866–81.

8. High Seas Fisheries

Fromentin, J. M. "Atlantic Bluefin." Chapter 2.1.5 in *ICCAT Field Manual*, 2006. http://www.iccat.int/Documents/SCRS/Manual/CH2/2_1_5_BFT_ENG.pdf.

Fromentin, J. M., and C. Ravier. "The East Atlantic and Mediterranean Bluefin Tuna Stock: Looking for Sustainability in a Context of Large Uncertainties and Strong Political Pressures." *Bulletin of Marine Science* 76, no. 2 (2005): 353–61.

MacKenzie, B. R., H. Mosegaard, and A. A. Rosenberg. "Impending Collapse of Bluefin Tuna in the Northeast Atlantic and Mediterranean." *Conservation Letters* 2 (2009): 25–34.

McAllister, M. K. and T. Carruthers. "Stock Assessment and Projections for Western Atlantic Bluefin Tuna Using a BSP and other SRA Methodology." *Collective Volume of Scientific Papers ICCAT* 62, no. 4 (2007): 1206–70.

9. Deepwater Fisheries

Branch, T. A. "A Review of Orange Roughy (Hoplostethus atlanticus) Fisheries, Estimation Methods, Biology and Stock Structure." *South African Journal of Marine Science—Suid-Afrikaanse Tydskrif Vir Seewetenskap* 23 (2001): 181–203.

Clark, M. "Are Deepwater Fisheries Sustainable? The Example of Orange Roughy (Hoplostethus atlanticus) in New Zealand." *Fisheries Research* 51, nos. 2–3 (2001): 123–35.

Francis, R. I. C. C., and M. R. Clark. "Sustainability Issues for Orange Roughy Fisheries." *Bulletin of Marine Science* 76, no. 2 (2005): 337–51.

Greenpeace Web site http://www.greenpeace.org/new-zealand/en/reports/New-Zealand-orange-roughy/ downloaded 4 January 2010

Hilborn, R., J. Annala, and D. S. Holland. "The Cost of Overfishing and Management Strategies for New Fisheries on Slow-growing Fish: Orange Roughy (Hoplostethus atlanticus) in New Zealand." *Canadian Journal of Fisheries and Aquatic Sciences* 63, no. 10 (2006): 2149–53.

World Wildlife Fund Web site http://www.worldwildlife.org/who/media/press/2003/WWFPresitem676.html downloaded 4 January 2010

10. Recreational Fisheries

American Sportfish Association Web site http://www.asafishing.org/images/statistics/resources/Sportfishing%20in%20America%20Rev.%207%2008.pdf downloaded 7 January 2010

Cook, Steven J., and Ian G. Cowx. "Contrasting Recreational and Commercial Fishing." *Biological Conservation* 128, no.1 (2006): 93–108.

Pitcher, Tony J., and Chuck Hollingworth, eds. *Recreation Fisheries: Ecological, Economic, and Social Evaluations.* Hoboken, NJ: Wiley-Blackwell, 2002.

11. Chilean Artisanal Fisheries

Castilla, J. C., and M. Fernández. "Small-scale Benthic Fishes in Chile: On Co-management and Sustainable Use of Benthic Invertebrates." *Ecological Applications* 8 (Supplement) (1998): S124–S132.

Castilla, J. C., and O. Defeo. "Latin American Benthic Shellfisheries: Emphasis on Co-management and Experimental Practices." *Reviews in Fish Biology and Fisheries* 11 (2001): 1–30.

Castilla, J. C., P. Manriquez, J. Alvarado, A. Rosson, C. Pino, C. Espoz, R. Soto, D. Oliva, and O. Defeo. "Artisanal 'Caletas' as Units of Production and Comanagers of Benthic Invertebrates in Chile." Proceedings of the North Pacific Symposium on Invertebrate Stock Assessment and Management. *Canadian Special Publication of Fisheries and Aquatic Sciences* 125 (1998): 407–13.

Gelcich, S., T. P. Hughes, P. Olsson, C. Folke, O. Defeo, M. Fernandez, S. Foale, L. H. Gunderson, C. Rodriguez-Sickert, M. Scheffer, R. S. Steneck, and J. C. Castilla. "Navigating Transformations in Governance of Chilean Marine Coastal Resources." *Proceedings of the National Academy of Sciences of the United States of America* 107 (2010): 16794–99.

Gutierrez, N. L., R. Hilborn, and O. Defeo. "Leadership, Social Capital and Incentives Promote Successful Fisheries." *Nature* 470 (2011): 386–89.

Orensanz, J. M., and A. M. Parma. "Chile—Territorial Use Rights: Successful Experiment?" *Samudra* 55 (2010): 42–46.

San Martín, G., A. M. Parma, and J. M. Orensanz. "The Chilean Experience with Territorial Use Rights in Fisheries." In *Handbook of Marine Fisheries Conservation and Management,* ed. R. Q. Grafton, R. Hilborn, D. Squires, M. Tait, and M. Williams, 324–37. Oxford: Oxford University Press, 2009.

Townsend, R., and R. Shotton. "Case Studies in Fisheries Self-governance." FAO fisheries technical paper 504. Rome: Food and Agriculture Organization of the United Nations, 2008.

12. Illegal Fishing

Agnew, D., J. Pearce, G. Pramod, T. Peatman, R. Watson, J. R. Beddington, and T. J. Pitcher. "Estimating the Worldwide Extent of Illegal Fishing." (2009). PLoS ONE e4570. doi:10.1371/journal.pone.0004570.

Knecht, G. B. *Hooked: Pirates, Poaching and the Perfect Fish.* Emmaus, PA: Rodale Press, 2006.

Lack, M. *Continuing CCAMLR's Fight against IUU Fishing for Toothfish.* WWF Australia and TRAFFIC International, 2008. http://www.wwf.or.jp/activities/upfiles/08-Continuing_CCAMLRs_Fight.pdf.

13. Trawling Impacts on Ecosystems

Collie, J. S., S. J. Hall, M. J. Kaiser, and I. R. Poiner. "A Quantitative Analysis of Fishing Impacts on Shelf-sea Benthos." *Journal of Animal Ecology* 69, no. 5 (2000): 785–98.

Hiddink, J. G., S. Jennings, M. J. Kaiser, A. M. Queirós, D. E. Duplisea, and G. J. Piet. "Cumulative Impacts of Seabed Trawl Disturbance on Benthic Biomass, Production, and Species Richness in Different Habitats." *Canadian Journal of Fisheries and Aquatic Sciences* 63, no. 4 (2006): 721–36.

Jennings, S., and M. J. Kaiser. "The Effects of Fishing on Marine Ecosystems." *Advances in Marine Biology* 34 (1998): 201–352.

National Research Council. *Effects of Trawling and Dredging on Seafloor Habitat*. Committee on Ecosystem Effects of Fishing: Phase 1–Effects of Bottom Trawling on Seafloor Habitats. Ocean Studies Board, Division of Earth and Life Sciences, National Research Council. Washington, DC: National Academy Press, 2002.

Pitcher, C. R., C. Y. Burridge, T. J. Wassenberg, B. J. Hill, and I. R. Poiner. "A Large Scale BACI Experiment to Test the Effects of Prawn Trawling on Seabed Biota in a Closed Area of the Great Barrier Reef Marine Park, Australia." *Fisheries Research* 99, no. 3 (2009): 168–83.

Sainsbury, K. J. "Application of an Experimental Approach to Management of a Tropical Multispecies Fishery with Highly Uncertain Dynamics." *Ecology* 193 (1991): 301–20.

Sainsbury, K. J., R. A. Campbell, R. Lindholm, and A. W. Whitlaw. "Experimental Management of an Australian Multispecies Fishery: Examining the Possibility of Trawl-induced Habitat Modification." In *Global Trends. Fisheries Management,* ed. E. K. Pikitch, D. D. Huppert and M. P. Sissenwine, 107–12. Seattle: American Fisheries Society, 1997.

Watling, L., and E. A. Norse. "Disturbance of the Seabed by Mobile Fishing Gear: A Comparison to Forest Clearcutting." *Conservation Biology* 12, no. 6 (1998): 1180–97.

14. Marine Protected Areas

Great Barrier Reef Marine Park Authority. *Outlook Report 2009.* http://www.gbrmpa.gov.au/corp_site/about_us/great_barrier_reef_outlook _report .

Hilborn, R., K. Stokes, J. J. Maguire, T. Smith, L. W. Botsford, M. Mangel, J. Orensanz, A. Parma, J. Rice, J. Bell, K. L. Cochrane, S. Garcia, S. J. Hall, G. P. Kirkwood, K. Sainsbury, G. Stefansson, and C. Walters. "When Can Marine Reserves Improve Fisheries Management?" *Ocean Coastal Management* 47 (2004): 197–205.

Jennings, S. "Role of Marine Protected Areas in Environmental Management." *ICES Journal of Marine Science* 66 (2009): 16–21.

National Research Council. *Marine Protected Areas: Tools for Sustaining Ocean Ecosystems.* Washington, DC: National Academy Press, 2001.

Norse, E. A., C. B. Grimes, S. Ralston, R. Hilborn, J. C. Castilla, S. R. Palumbi, D. Fraser, and P. Kareiva. "Marine Reserves: The Best Option for Our Oceans?" *Frontiers in Ecology and Evolution* 1 (2003): 495–502.

Wood, L. J., L. Fish, J. Laughren, and D. Pauly. "Assessing Progress towards Global Marine Protection Targets: Shortfalls in Information and Action." *Oryx* 42 (2008): 340–51.

15. Ecosystem Impacts of Fishing

Carpenter, S. R. and J. F. Kitchell, eds. *The Trophic Cascade in Lakes.* Cambridge: Cambridge University Press, 1993.

Carpenter, S. R., J. J. Cole, J. F. Kitchell, and M. L. Pace. "Trophic Cascades in Lakes: Lessons and Prospects." In *Trophic Cascades*, ed. J. Terborgh and J. Estes, 55–70. Washington, DC: Island Press, 2010.

Pikitch, E. K., C. Santora, E. A. Babcock, A. Bakun, R. Bonfil, D. O. Conover, P. Dayton, P. Doukakis, D. Fluharty, B. Heneman, E. D. Houde , J. Link, P. A. Livingston, M. Mangel, M. K. McAllister, J. Pope, and K. J. Sainsbury. "Ecosystem-Based Fishery Management." *Science* 305 (2004): 346–47.

16. Status of Overfishing around the World

Branch, T. A., O. P. Jensen, D. Ricard, Y. Ye, and R. Hilborn. "Contrasting Global Trends in Marine Fishery Status Obtained from Catches and from Stock Assessments." *Conservation Biology* 25 (2011): 777–86.

Hilborn, R., T. A. Branch, B. Ernst, A. Magnusson, C. V. Minte-Vera, M. D. Scheuerell, and J. L. Valero. "State of the World's Fisheries." *Annual Review of Environment and Resources* 28 (2003): 359–99.

Hutchings, J. A., C. Minto, D. Ricard, J. K. Baum, and O. P. Jensen. "Trends in Abundance of Marine Fishes." *Canadian Journal of Fisheries and Aquatic Sciences* 67 (2010): 1205–10.

Jackson, J. B. C., M. X. Kirby, W. H. Berger, K. A. Bjorndal, L. W. Botsford, B. J. Bourque, R. H. Bradbury, R. Cooke, J. Erlandson, J. A. Estes, T. P. Hughes, S. Kidwell, C. B. Lange, H. S. Lenihan, J. M. Pandolfi, C. H. Peterson, R. S. Steneck, M. J. Tegner, and R. R. Warner. "Historical Overfishing and the Recent Collapse of Coastal Ecosystems." *Science* 293 (2001): 629–38.

Lotze, H. K., H. S. Lenihan, B. J. Bourque, R. H. Bradbury, R. G. Cooke, M. C. Kay, S. M. Kidwell, M. X. Kirby, C. H. Peterson, and J. B. C. Jackson. "Depletion, Degradation, and Recovery Potential of Estuaries and Coastal Seas." *Science* 312 (2006): 1806–09.

Worm, B., R. Hilborn, J. K. Baum, T. A. Branch, J. S. Collie, C. Costello, M. J. Fogarty, E. A. Fulton, J. A. Hutchings, S. Jennings, O. P. Jensen, H. K. Lotze, P. M. Mace, T. R. McClanahan, C. Minto, S. R. Palumbi, A. Parma, D. Ricard, A. A. Rosenberg, R. Watson, and D. Zeller. "Rebuilding Global Fisheries." *Science* 325 (2009): 578–85.

INDEX